Elements

SILICON

Si

Atlantic Europe Publishing

How to use this book

This book has been carefully developed to help you understand the chemistry of the elements. In it you will find a systematic and comprehensive coverage of the basic qualities of each element. Each two-page entry contains information at various levels of technical content and language, along with definitions of useful technical terms, as shown in the thumbnail diagram to the right. There is a comprehensive glossary of technical terms at the back of the book, along with an extensive index, key facts, an explanation of the Periodic Table, and a description of how to interpret chemical equations.

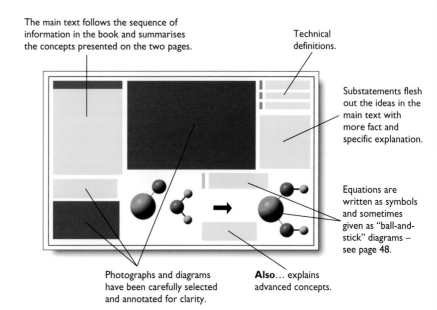

The main text follows the sequence of information in the book and summarises the concepts presented on the two pages.

Technical definitions.

Substatements flesh out the ideas in the main text with more fact and specific explanation.

Equations are written as symbols and sometimes given as "ball-and-stick" diagrams – see page 48.

Photographs and diagrams have been carefully selected and annotated for clarity.

Also... explains advanced concepts.

An Atlantic Europe Publishing Book

Author
Brian Knapp, BSc, PhD
Project consultant
Keith B. Walshaw, MA, BSc, DPhil
(Head of Chemistry, Leighton Park School)
Industrial consultant
Jack Brettle, BSc, PhD (Chief Research Scientist, Pilkington plc)
Art Director
Duncan McCrae, BSc
Editor
Elizabeth Walker, BA
Special photography
Ian Gledhill
Illustrations
David Woodroffe
Electronic page make-up
Julie James Graphic Design
Designed and produced by
EARTHSCAPE EDITIONS
Print consultants
Landmark Production Consultants Ltd
Reproduced by
Leo Reprographics
Printed and bound by
Paramount Printing Company Ltd

Suggested cataloguing location
Knapp, Brian
   Silicon
   ISBN 1 869860 34 9
   – *Elements* series
540

Acknowledgements
The publishers would like to thank the following for their kind help and advice: *Rolls-Royce plc, Janet McCrae, David Newell and Dr Stephen Codrington.*

Picture credits
All photographs are from the **Earthscape Editions** photolibrary except the following:
(c=centre t=top b=bottom l=left r=right)
Ian Gledhill 45; courtesy of **Rolls-Royce plc** 30cr, 31t, 31b; **NASA** 32/33c and **ZEFA** 4/5b, 14b, 35tl, 36/37, 40/41t.

Front cover: Ceramics such as this crucible and the clay pipe it sits on are useful in laboratory chemistry. They are relatively cheap to manufacture and can withstand high temperatures without melting.
Title page: A beautiful piece of hand blown Venetian glass.

First published in 1996 by
Atlantic Europe Publishing Company Limited, Greys Court Farm,
Greys Court, Henley-on-Thames, Oxon, RG9 4PG, UK.

Copyright © 1996
Atlantic Europe Publishing Company Limited
Reprinted in 1997

This product is manufactured from sustainable managed forests. For every tree cut down at least one more is planted.

The demonstrations described or illustrated in this book are not for replication. The Publisher cannot accept any responsibility for any accidents or injuries that may result from conducting the experiments described or illustrated in this book.

Contents

Introduction

An element is a substance that cannot be broken down into a simpler substance by any known means. Each of the 92 naturally occurring elements is therefore one of the fundamental materials from which everything in the Universe is made. This book is about silicon.

Silicon

There is more silicon in the Earth's crust than any other element except oxygen, and yet we are usually completely unaware of its presence. In part this is because silicon has formed very stable compounds and in part because it is so common we take it for granted.

Whenever you pick up a rock, the chances are that silicon, along with oxygen, will make up most of it. These elements are combined as the mineral called silica. Only limestones, coals and salt beds are relatively free of silica. All sandstones, shales, and volcanic rocks contain it, as do the majority of soil particles.

We use silica to make concrete, bricks and glass, which are among the most important building materials of our world. We also use a grit of silica for "sanding" rough wooden and plastic surfaces smooth. In fact, the silicates, those minerals in which silicon is a main component, also make up some of the world's hardest, most beautiful and sought-after minerals. Emerald and aquamarine are silicate rocks, although most of us think of them first as gems or precious stones.

In recent years scientists have found far more uses for what was, in the past, thought of as a rather unpromising element. The world's computers and all computer controlled equipment have, at their heart, "silicon chips", crystals of silica that are so pure that only one impure atom in a billion contaminates them beyond use. Crystals of silicon and chip slices are pictured on this page.

▲ This shows some crystals and parts of a wafer of high purity silicon, the element used to make silicon chips.

Silicate garden

Silica does not readily react with many substances. We rely on this property in our everyday lives. For example, we use bricks containing silica to build our houses and glass containing silica for window panes. Think what would happen if bricks changed whenever other substances touched them! But under some conditions even silica can be made to produce some spectacular effects, such as those shown here.

Water glass (sodium silicate) is a soluble compound that results from the reaction of silica with sodium carbonate. This reaction takes place at very high temperatures (1400°C).

This picture shows a "chemical garden" made using sodium silicate solution in water. Several crystals of different metal salts have been dropped into the bowl. As each crystal begins to dissolve it reacts with the sodium silicate and forms a bubble of material with a silicate "skin". For example, the skin of the bubble around the cobalt chloride crystal, is cobalt silicate. This skin is tough, but it can let water pass through. As water passes from the water glass into the cobalt silicate it causes the pressure to increase inside the bubble until it bursts at the top. The reaction starts again and a new bubble forms.

By repeated bursting and forming of bubbles, the columns grow taller.

Blue copper silicate

Yellow chromium silicate

Green nickel silicate

Brown manganese silicate

Blue cobalt silicate

The world of silicate minerals

Silicon occurs in about one-third of the minerals in the Earth's crust. Clearly, therefore, silicon is a very important element.

Over one thousand different minerals have been recognised as silicates. Silicon is nearly always combined with oxygen to make the compound silica (SiO_2), which is the building block of so many of the world's minerals and rocks.

In terms of volume, silicates make up about 93% of the Earth's crust. Minerals that comprise only silica molecules are called silica; those in which silica combines with other metal elements are called silicates.

Silicate compounds are not very reactive. This has the advantage of making the Earth's surface a stable place on which to live. This is not to say that silicon makes uninteresting compounds; far from it. But nature, rather than people, is the chemist here, forming and reforming a remarkable variety of structures and giving us the beautiful minerals we see around us.

Silica building blocks

The variety in silicates is accounted for by the way the silica units can group, or cluster, together. Just as a huge variety of plastics can be produced from simple components that cluster, or polymerise, together, so silica units can group together into chains, rings, sheets and framework patterns. The importance of these structures will be shown on these and some following pages.

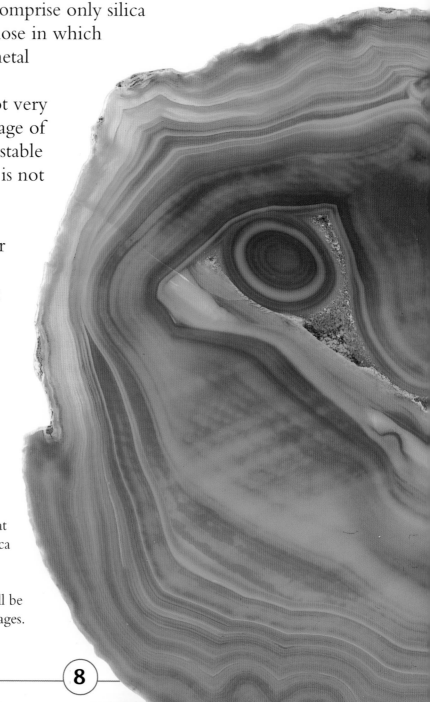

Silicate groups

There are many well-known mineral groups, each with its own unique pattern of silica molecules. For example, in the mineral quartz (page 10) the molecules build up in frameworks. Other silicates, for example, the group containing emerald, are built from rings of silica molecules. You will find the silicates presented in these different groups on the following pages.

mineral: a solid substance made of just one element or chemical compound. Calcite is a mineral because it consists only of calcium carbonate, halite is a mineral because it contains only sodium chloride, quartz is a mineral because it consists of only silicon dioxide.

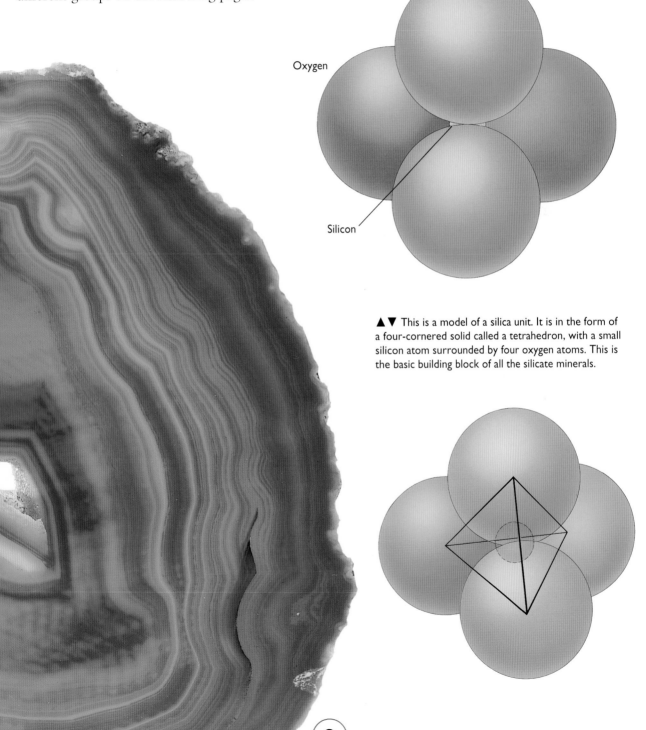

Oxygen

Silicon

▲ ▼ This is a model of a silica unit. It is in the form of a four-cornered solid called a tetrahedron, with a small silicon atom surrounded by four oxygen atoms. This is the basic building block of all the silicate minerals.

Quartz and feldspar

Quartz, also known as "rock crystal", is the most commonly found silica mineral, being made almost entirely from silicon and oxygen atoms.

Pure quartz is colourless and transparent, but it is often contaminated with impurities that produce a range of colours. One of the most common coloured quartzes is called "smoky quartz".

Quartz is a very hard mineral and cannot be scratched with a knife. It forms into long, six-sided (hexagonal) crystals.

Quartz is an example of a mineral made of silica molecules that extend outwards in every direction to make a framework. No other elements are involved in building the frame. This makes quartz a particularly stable mineral.

The occurrence of crystalline silica
Free silica is commonly found as either large crystals or small grains. Quartz is also common as noncrystalline silica filling in veins and other fractures in rock.

▲ The structure of silica is very similar to that of diamond, with the silicon atoms packed tightly. Four large oxygen atoms lie at the corners of a tetrahedron, almost entirely surrounding a smaller silicon atom (see the diagram on page 9). The electrical charges between the silicon and the oxygen atoms balance exactly, forming strong bonds. This structure makes minerals like quartz very stable and unreactive.

◄ In this crystal of quartz, the hexagonal (six-sided) crystals can be seen. These crystals are about six centimetres long; some quartz crystals grow to a huge size and can weigh several tonnes.

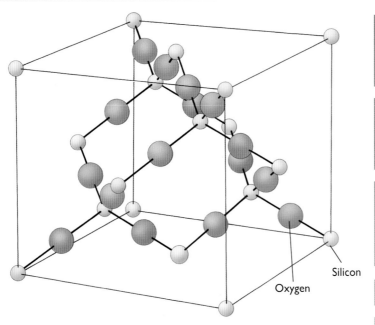

▲ The arrangement of silica molecules in quartz is in the form of a kind of interlocking building frame that stretches out in every direction. This is why the mineral is hard and brittle.

crystal: a substance that has grown freely so that it can develop external faces. Compare with crystalline, where the atoms were not free to form individual crystals and amorphous where the atoms are arranged irregularly.

igneous rock: a rock that has solidified from molten rock, either volcanic lava on the Earth's surface or magma deep underground. In either case the rock develops a network of interlocking crystals.

mineral: a solid substance made of just one element or chemical compound. Calcite is a mineral because it consists only of calcium carbonate, halite is a mineral because it contains only sodium chloride, quartz is a mineral because it consists of only silicon dioxide.

molecule: a group of two or more atoms held together by chemical bonds.

translucent: almost transparent.

Common properties of the silicates

Although there are over a thousand different minerals in this group, you might be surprised to learn that there are many common characteristics. In general, most silicates are hard and do not dissolve in water; most do not even dissolve in acids. The crystals of silicate minerals tend to look glassy, and most of the crystals are transparent or translucent.

Feldspars

Feldspar minerals are found in most igneous (volcanic) rocks. They are silicates in which some of the silicon atoms have been replaced by aluminium atoms. This is because aluminium is a small atom and can also balance out the charges on the oxygen atoms. When this happens, however, some "holes" are left in the structure, in which potassium, sodium or calcium atoms can also fit. This produces a variety of feldspars.

Feldspar crystals are opaque and either pink, grey or white. You can see them clearly in the granite sample shown here.

▲ A piece of granite. The grey, glassy-looking crystals are quartz. The inset picture shows some granite scenery in the Mojave Desert, California, USA.

Quartz as a gemstone

People have often treasured the brightest and most sparkling of the Earth's rocks and minerals. The total number of minerals in the world is thought to be about three thousand, but only one hundred are thought of as gemstones, and of these only thirteen are really common.

Most gemstones are silicate minerals, either in their uncrystalline form, such as jade, or as crystals, such as emerald.

Some less valuable silicate minerals are often mistaken for gemstones by the inexperienced. One such mineral is zirconium, used as a substitute for diamond, which is actually produced synthetically. Emerald can now also be made industrially. The crystals of synthetic emeralds are usually of superior quality to those found in nature.

Quartz as a gemstone

Quartz, silicon dioxide, is an example of a silicate, a compound containing silicon and oxygen. But pure quartz is by no means the only example of a silicate. Most of the world's gemstones are also silicates.

Coloured quartz is known by a variety of names. Violet-coloured quartz is called amethyst, black quartz is smoky quartz, and yellow colouring produces a form of quartz called citrine.

Amethyst

Amethyst is a form of quartz that has been contaminated with iron, manganese and carbon and has taken on a violet hue. It changes to orange–yellow citrine on heating.

◀ Amethyst.

▼ Smoky quartz, a dark brown form of quartz.

gemstone: a wide range of minerals valued by people, both as crystals (such as emerald) and as decorative stones (such as agate). There is no single chemical formula for a gemstone.

precipitate: tiny solid particles formed as a result of a chemical reaction between two liquids or gases.

solution: a mixture of a liquid and at least one other substance (e.g. salt water). Mixtures can be separated out by physical means, for example by evaporation and cooling.

▼ Cat's-eye gets its name from the luminous, reflective bands that resemble the slit pupil of a cat. The reflective area contains asbestos fibres and the green colour comes from the copper oxides in the silica structure.

▲ Tiger's-eye shows banding properties that are similar to Cat's-eye, but the reddish-brown colour comes from iron oxides.

Also...
The main gemstones that are not silicates are diamond (made from carbon) and ruby (made from aluminium oxide).

▼ Agate.

Tiger's-eye, cat's-eye and agate

These banded forms of silica are considered to be gemstones. They enter cavities inside rocks as silica-rich solutions. These then cool, and the silica is precipitated on the inside of the cavity. Typically, such cavities form inside lavas and other igneous rocks when they are cooling. Jasper (red), chalcedony (grey) and carnellian (red) are all forms of the same kind of deposition.

Other forms of quartz

Of the many forms of quartz, opal has no crystals at all, while many others, such as obsidian and flint, have crystals that can only be seen with a microscope. These are known as "microcrystalline" minerals.

Flint arrowheads

Cutting edge

Flint

Flint is a form of microcrystalline silica. This silica was originally carried from silicate rocks by percolating waters and reprecipitated elsewhere. (Agate, shown on page 12, is another form of silica precipitate, but it has developed banding that makes it an attractive gemstone.)

Flint is common in chalk rocks and nodules of flint were once split to make axes and arrowheads. Flint and obsidian were the main axe-making materials of the Stone Age.

Notice that, because flint is a variety of framework silicates, it does not break along any particular surface and so can be sculpted easily. Because it is brittle, however, it can only be sculpted into rough shapes.

◀ Flint is very brittle and breaks with a conchoidal (curved) fracture. Skilled crafts people could use one flint to split another. This was called napping. The objective was to split off flakes that could be used as arrowheads, while creating a hand axe that had a sharp cutting edge or a point.

▶ This opal shows a characteristic pearlescent lustre.

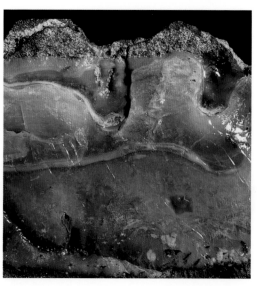

Opal

The one form of quartz that has no crystals at all is called opal. It forms as a precipitate from a solution of water containing silica. In fact, opal stones contain water molecules. As a result it is less dense than other forms of quartz. Opal also has a different lustre to other forms of quartz, being more like that of a pearl.

glass: a transparent silicate without any crystal growth. It has a glassy lustre and breaks with a curved fracture. Note that some minerals have all these features and are therefore natural glasses. Household glass is a synthetic silicate.

lustre: the shininess of a substance.

solution: a mixture of a liquid and at least one other substance (e.g. salt water). Mixtures can be separated out by physical means, for example by evaporation and cooling.

Obsidian

Obsidian is called natural glass. It is a form of silica with very small crystals, usually formed from molten lava, where cooling was so rapid that large crystals did not have time to develop. Lava that erupts under the sea often has this property.

Like flint, obsidian breaks with a conchoidal fracture. It has a vitreous lustre, and is usually black. An example of the fracture is shown here.

▼ A piece of natural volcanic glass, obsidian, showing conchoidal fracturing.

► This sculpture is made from obsidian, a framework silicate that does not have any special crystal shape and so is easy to carve. It is in the Olmec (Central American) style.

Quartzite veins

Silica rarely makes up a rock on its own. Rather, it is found as veins between other rocks. Occasionally the vein is thick enough to form a band of rock, when it is known as quartzite.

The presence of quartz is also an indicator that, in the geological past, conditions were hot enough for silica to be melted. Geologists know that in similar conditions metals and valuable minerals are likely to be molten. So quartzite veins are a tell-tale sign of the presence of everything from tin to gold.

► This is a quartzite vein in the Bendigo goldfield of Australia. Quartz forms from a hot liquid only after other silicate minerals have formed, because it has the lowest melting point of any silicate.

▲ Canyon of the Gunnison River, Colorado, USA, showing well developed quartzite veins that have been exposed by river erosion.

vein: a mineral deposit different from, and usually cutting across, the surrounding rocks. Most mineral and metal-bearing veins are deposits filling fractures. The veins were filled by hot, mineral-rich waters rising upwards from liquid volcanic magma. They are important sources of many metals, such as silver and gold, and also minerals such as gemstones. Veins are usually narrow, and were best suited to hand-mining. They are less exploited in the modern machine age.

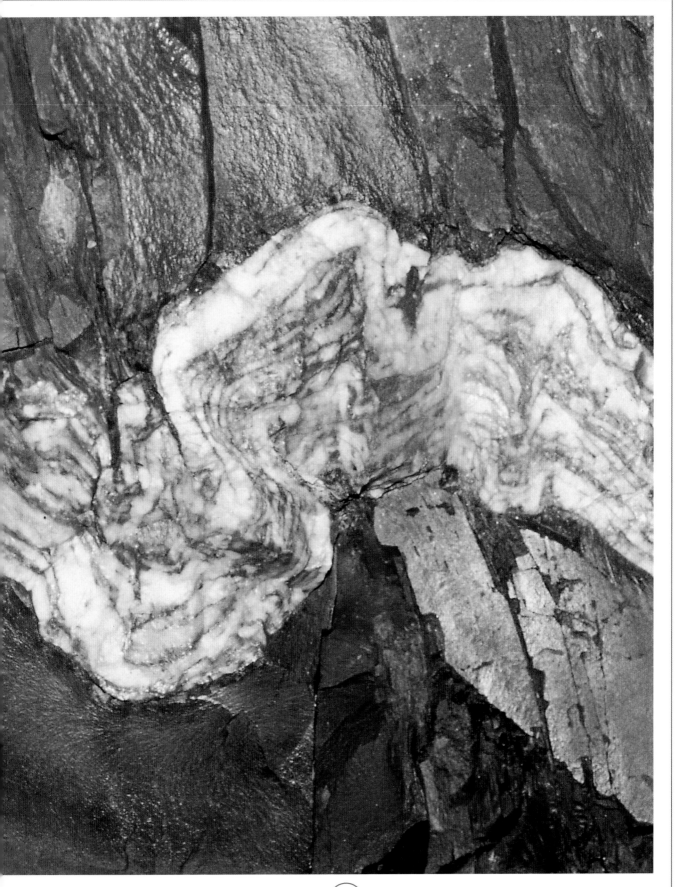

Sandstone

Silicates, and in particular quartz, are among the least reactive substances and so are most resistant to weathering. The quartz grains formed in igneous rocks are released as the other minerals are weathered away. As they roll about in rivers or are tossed about by coastal currents, the edges of the quartz grains are smoothed down and become more rounded in shape. However, they do not change chemically and so remain to form the skeleton of many later rocks. This is why sandstones are among the most common rocks.

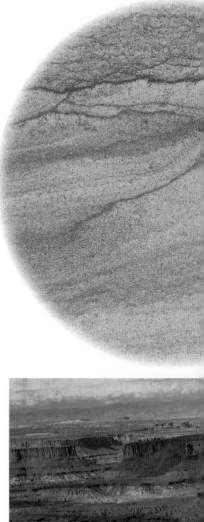

Sandstones are grains of quartz and other small, unweathered rock fragments cemented together by another mineral. Geologists call such rocks sedimentary rocks because they are formed of layers of rocks fragments (sediments) that have settled out onto the ocean floor or some other place where, over time, they become "cemented" together. In this process the gaps between the grains become filled in.

The main forms of natural sandstone "cement" are iron oxide and calcite. Iron oxide cement is red, orange or yellow and stains the outside of the sand grains, as well as sometimes forming thin bands of highly concentrated material. The iron staining helps to highlight patterns in the sandstone, making it easier to see in what conditions the sand grains were laid down. For example, the sandstone shown here was formed from beach deposits and the ripples of small waves can still be seen.

Calcite (calcium carbonate) is a white cement that does not stain the sand grains. As a result, sandstones cemented with calcite are light-coloured. Calcium carbonate is much more likely to become weathered than iron oxide, so many calcite-cemented rocks crumble relatively easily when exposed to wet weather.

◄ The sandstone in this rock sample was deposited in a shallow sea. The ripples are still visible.

▼ Sand dunes in Death Valley, California, USA.

weathering: the slow natural processes that break down rocks and reduce them to small fragments either by mechanical or chemical means.

◄ Iron oxide cements are very durable because they are resistant to weathering. The high red sandstone cliffs of the Canyonlands of the southwest United States, for example, including the cliffs of the Grand Canyon, bear witness to this.

Garnets, zircon and olivine

In these silicates the silica molecules are held together by metal atoms. This influences the shape of the crystals in each mineral.

The minerals in this group form at very high temperatures and are only in rocks associated with volcanoes and mountain-building. Two of the more easily recognised are garnet and zircon. However, the most widespread of these minerals is called olivine, a mineral found in all dark-coloured igneous rocks.

Garnet

Garnets are common minerals, varying in colour from green and yellow to deep, rich red. All crystal varieties within the group have a cubic shape.

Garnets are combinations of iron and silica with various combinations of calcium, aluminium and manganese. The various metals are responsible for the variety of colours.

Garnets are all hard and have a glassy lustre. They are formed in granites but are especially common in rocks that have been subjected to considerable heat and pressure (known as metamorphic rocks), where they form crystals much harder than the surrounding rock. Weathering of rocks causes the garnets to protrude and become easy to spot. One of the most common of these rocks is a metamorphic rock called schist.

▶ Garnets are often found as single cubic crystals with a characteristic opaque pink colouration. Garnetiferous mica–schist is a common metamorphic rock. The majority of the rock is made of mica crystals, which gives it its sheen; individual small transparent red garnet crystals grow in the mica.

Zircon

Zircon is the more common name for brilliant sparkling diamond-like crystals of zirconium silicate.

It often forms pyramid-headed crystals in nature. Zircon can be found as tiny crystals in many igneous rocks, where it may vary in colour from brown to green or be colourless. Zircon is very resistant to weathering and so is also found in sediments such as sandstone, where it often makes up dark grains among the pale grains of quartz.

The transparent form is called "Matura diamond". It has a structure of atoms similar to real diamond. Synthetic zircon is used in jewellery as a substitute for diamonds (known commonly as CZs, cubic zirconia).

igneous rock: a rock that has solidified from molten rock, either volcanic lava on the Earth's surface or magma deep underground. In either case the rock develops a network of interlocking crystals.

metamorphic rock: formed either from igneous or sedimentary rocks, by heat and/or pressure. Metamorphic rocks form deep inside mountains during periods of mountain building. They result from the remelting of rocks during which process crystals are able to grow. Metamorphic rocks often show signs of banding and partial melting.

▲ Cubic zirconia is used in inexpensive jewellery.

Olivine

Olivines are greenish minerals found in many igneous rocks. They are silicates held together with iron and magnesium. They do not usually form crystals, but instead form grains of translucent, brittle mineral.

They are minerals associated with very high temperature and pressure and are more common in rocks that have been made from materials deep within the earth. The diamond pipes of Kimberley, South Africa, consist mainly of olivine.

▲ Olivine-rich rock showing the characteristic green colouration.

◄ Kyanite can form beautiful blue blade-shaped crystals, and is one of the major mineral resources of India.

Beryl and the ring silicates

Sometimes silica molecules form into rings, usually in groups of six. This is not a very common occurrence, but it produces some of the world's most spectacular minerals. The common form is called beryl, but the gemstone forms are emerald and aquamarine.

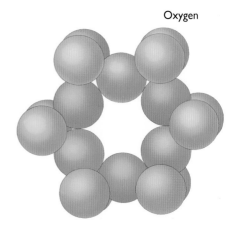

Oxygen

Beryl

Beryl is made of columns of hexagonal crystals. The glassy green mineral is often found in large pieces, sometimes several metres long. They are formed by rings of six silica molecules stacked one above the other in sheets. Beryllium atoms help to hold the sheets together.

Beryl is commonly found in granites, along with tourmaline.

▶ Beryl.

▲ Emerald.

Emerald

This is a gemstone with the same chemical structure as beryl, but with the deep clear green colour being produced by chromium atoms. These atoms make emeralds softer than other gemstones, for example, ruby.

▶ Aquamarine.

igneous rock: a rock that has solidified from molten rock, either volcanic lava on the Earth's surface or magma deep underground. In either case the rock develops a network of interlocking crystals.

metamorphic rock: formed either from igneous or sedimentary rocks, by heat and/or pressure. Metamorphic rocks form deep inside mountains during periods of mountain building. They result from the remelting of rocks during which process crystals are able to grow. Metamorphic rocks often show signs of banding and partial melting.

Aquamarine
Aquamarine is a clear form of beryl that is regarded as a gemstone. The blue–green colour comes from the elements chromium and iron.

◀ Tourmaline.

Tourmaline
Tourmaline is a silicate compounded with boron. (This is the same chemical unit that makes some glass heat-resistant.) It commonly forms black needle-like crystals in igneous rocks, especially granites. The faces of the crystals have scratches (striations) on them that make the surface look like a piece of a record. The crystals are hard and have a glassy lustre. One form of tourmaline becomes electrically charged when its temperature is changed.

Jade and the chain silicates

It is more common for silica molecules to form long chains, similar to the polymerisation that allows simple organic molecules to be formed into plastics.

Chain silicates are among the most common minerals. They are very strong in the direction of the chain, but they are much weaker between chains. This means that they tend to break parallel to the chains.

Pyroxenes and amphiboles

The pyroxenes are an extensive group of minerals. Not only do they form in nature, but they also form in many industrial processes. For example, the slag produced during the refining of iron ore in a blast furnace often has the mineral composition of a pyroxene.

Pyroxenes are the major component of basaltic lava, the most common rock at the Earth's surface (it underlies all the world's oceans). Thus, pyroxenes though not widely known, are among the most common minerals.

The most frequently seen pyroxene mineral, augite, is dark green to black.

Amphibole has the same chemical composition as the pyroxenes, but it forms at a lower temperature. The most common mineral of this group is hornblende, a dark green to black mineral found in all basalts.

Because most basalts cool quickly, only small crystals form in them, so it is usually very difficult to distinguish between the pyroxenes and amphiboles, as both look like small, dark crystals.

Jade

Jade is a gemstone that is especially highly regarded in Asia. It is a combination of a chain of silica molecules with sodium and aluminium. The green colour is produced by atoms of iron.

Jade does not occur as crystals, but rather as large lumps of green mineral, which is what makes it suitable for carving. It is formed under conditions of high temperature and pressure deep within mountain systems and so is a mineral of metamorphic rocks.

▶ The main use for jade is in the art of the Far East. It was once used to make massive objects such as state coaches. The jade on this detail of a state coach weighs many tonnes.

▲ Augite.

24

igneous rock: a rock that has solidified from molten rock, either volcanic lava on the Earth's surface or magma deep underground. In either case the rock develops a network of interlocking crystals.

metamorphic rock: formed either from igneous or sedimentary rocks, by heat and/or pressure. Metamorphic rocks form deep inside mountains during periods of mountain building. They result from the remelting of rocks during which process crystals are able to grow. Metamorphic rocks often show signs of banding and partial melting.

polymerisation: a chemical reaction in which large number of similar molecules arrange themselves into large molecules, usually long chains. This process usually happens when there is a suitable catalyst present. For example, ethene reacts to form polythene in the presence of certain catalysts.

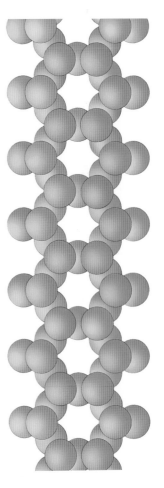

▲ A chain silicate.

Sheet minerals

In many cases, silica molecules form extensive sheets. Stacks of these sheets are connected together by metal ions; potassium and magnesium are among the most common. The way the sheets are connected together is very important, because it gives the minerals many of their properties.

Sheet silicates all break up into thin flakes. This can be important. Several of the sheet silicates are used as lubricants; for example, talcum powder is made from the mineral talc.

Water molecules can be absorbed between the sheets of many clay minerals. This is why pottery clays swell when they are wet and shrink when they are fired, and also why soils shrink and swell.

Kaolinite

There are many different clay minerals, but all are composed of sheets containing aluminium silicates. Kaolinite, a soft mineral that can absorb water, is the most common clay mineral in the world's soils. It is mined from concentrated deposits as china clay, and is important as the raw material for pottery and ceramics.

Kaolinite crystals are too small to be seen with an ordinary microscope.

Mica

Mica is the third most common group of minerals in the earth's crust, after quartz and the feldspars. It easily recognised as a silicate because it breaks up into thin, almost transparent sheets. The main elements, together with silicon and oxygen, are aluminium and potassium.

Varieties of mica are black biotite, and brown muscovite.

▼ Biotite is the black form of mica.

▶ Hexagonal crystals of vermiculite.

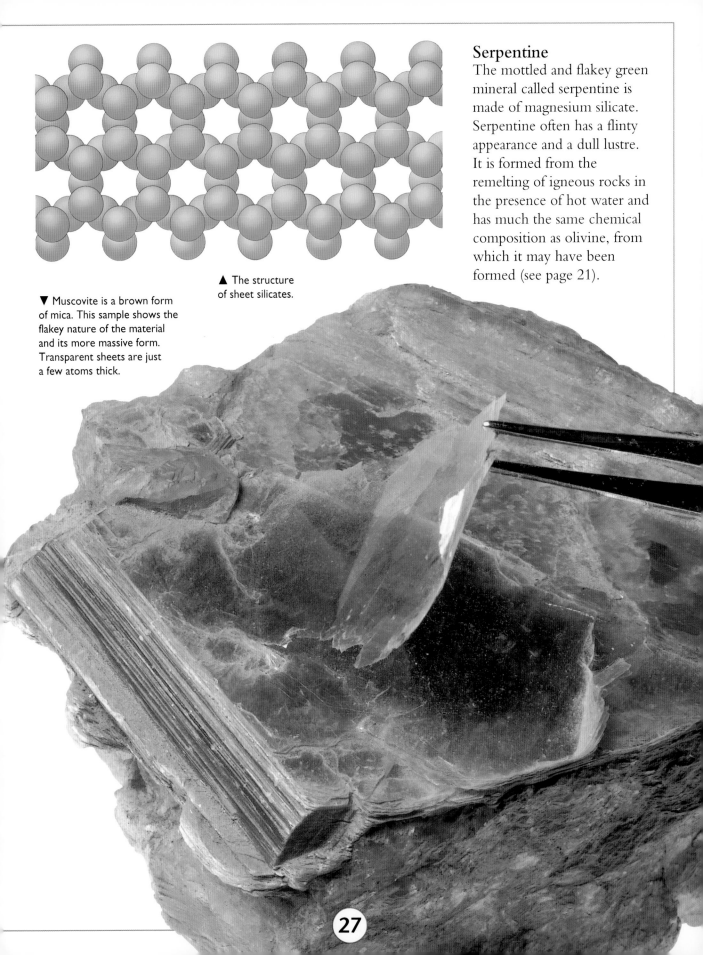

▲ The structure of sheet silicates.

Serpentine

The mottled and flakey green mineral called serpentine is made of magnesium silicate. Serpentine often has a flinty appearance and a dull lustre. It is formed from the remelting of igneous rocks in the presence of hot water and has much the same chemical composition as olivine, from which it may have been formed (see page 21).

▼ Muscovite is a brown form of mica. This sample shows the flakey nature of the material and its more massive form. Transparent sheets are just a few atoms thick.

Firing clay

When clay is heated, its properties change dramatically, and the soft sticky material is transformed into a hard brittle substance. This property is used for pottery and brick-making to make suitable shapes that can be set hard for use.

During intense heating, or firing, water that normally occurs between the sheets of the clay crystals is driven off, while quartz in the material melts. Thus the clay crystals become fused together by glassy silica.

China clay

China clay is a concentrated deposit of the mineral kaolinite. Kaolinite forms in large quantities when the hot fluids emanating from underground magma chambers intensely weather feldspar. Most china clay is white and can be used for porcelain and china. It is also used as a smooth coating on some paper.

▼ Air-dried pots being prepared for firing. The kiln can be seen in the background.

▲ Bricks are among the most important products made with sheet silicates.

ceramic: a material based on clay minerals which has been heated so that it has chemically hardened.

magma: the molten rock that forms a balloon-shaped chamber in the rock below a volcano. It is fed by rock moving upwards from below the crust.

oxidation: a reaction in which the oxidising agent loses electrons. (Note that oxidising agents do not have to contain oxygen.)

sintering: a process that happens at moderately high temperatures in some compounds. Grains begin to fuse together even through they do not melt. The most widespread example of sintering happens during the firing of clays to make ceramics.

▼ A kiln-fired porcelain pot.

The chemistry of firing

Ceramics are produced by firing (heating) clays in a kiln (oven). The air-dried clay objects put into the kiln may look and feel dry, but they still contain water molecules locked between the sheets of the clay minerals. This water is only lost when the clay is heated to a very high temperature.

At the same time, the high temperature causes an irreversible chemical reaction to occur. The clay changes into a new mineral (mullite) which has tiny needle-shaped crystals. This is what makes the ceramic strong. Also, minute grains of silica form and fuse together, a process called sintering.

These chemical changes may also cause the colour of the clay to change. If pure kaolinite is used, the ceramic becomes white. However, if the kaolinite contains or is mixed with impurities – compounds of metals such as chromium, iron or manganese – a coloured ceramic will be produced. This is because, during the heating process, the metal becomes oxidised. The oxides absorb some of the wavelengths of natural light, thus making the object appear coloured.

▼ Coloured ceramic tiles.

EQUATION: Chemical change during the firing of a ceramic

Kaolinite (clay) ⇨ *mullite (fired clay) + silica + water vapour*

$$3Al_2Si_2O_5(OH)_4(s) \quad \Rightarrow \quad Al_6Si_2O_{13}(s) \quad + \quad 4SiO_2(s) \quad + \quad 6H_2O(g)$$

Ceramics in engines of the future

Jet engines place the highest demands on materials. At the moment various different metals are used for turbine blades and other parts. However, these materials cannot meet the needs of the future, when engines will have to be lighter and yet more powerful. This is where new silicon compounds are likely to be used.

Silicon carbide is one such ceramic material. Components made with it can operate at temperatures up to 1600°C, well beyond the melting point of many metals.

The disadvantage of ceramics is their brittleness, so at the moment their use is limited. However, by making composite materials of fibre reinforced metals and ceramics, many of these problems can be overcome. Weight for weight, titanium metal reinforced with silicon carbide is twice as strong as a titanium alloy. Glass and glass–ceramic reinforced with silicon carbide fibres may also be used in the future for even stronger and lighter materials.

▼ Silicon carbide in mineral form is known as carborundum. It is very hard and widely used in abrasive powders.

▲ This is a computer model of how the temperature varies through a turbine blade. Making sure the ceramic composites can stand up to high temperatures and vibration is a target of research departments.

▼ The heat given out of the exhaust ports of this jet engine gives you a vivid idea of how important it is to have materials that can resist heat and also the force of the exhaust gases.

alloy: a mixture of a metal and various other elements.

ceramic: a material based on clay minerals which has been heated so that it has chemically hardened.

▼ Although most blades in turbine engines are still made of special steel, especially those containing the strong, light metal titanium, new technology is making it possible to design ceramic blades that will stand up to high temperatures better than metal blades. One such ceramic blade is shown in the centre of this picture; the other blades are made of metal.

Ceramic turbine blade

Titanium turbine blade

Silicones

Silicones are rubbery materials that contain silicon. Like all rubbery materials, the silicones have many molecules linked together in long chains by a process called polymerisation.

The nature of the polymer varies with the nature of additional water-repellent groups of atoms in the structure. Methyl groups are common substitutes because they make silicones waterproof. When methyl-containing silicones are used as waterproofers, some of the oxygen atoms of the silicones attach themselves to the fabric, leaving the methyl groups sticking out of the chains to repel water.

Silicones are quite unreactive compounds; thus, they can be used in a wide variety of environments where other materials (for example polymers based on petroleum) would decompose or react. Thus, silicone sealants are used around kitchen sinks, space shuttle fuel tanks and even for some human body implants.

► Silicone gasket rings are used in some of the important connectors between the fuel tanks and the engines of space rockets. They are able to withstand corrosive chemicals and high temperatures that are generated as the engines fire.

▲ Silicone can be used as a water sealant on clothing and umbrellas.

polymer: a compound that is made of long chains by combining molecules (called monomers) as repeating units. ("Poly" means many, "mer" means part).

polymerisation: a chemical reaction in which large number of similar molecules arrange themselves into large molecules, usually long chains. This process usually happens when there is a suitable catalyst present. For example, ethene reacts to form polythene in the presence of certain catalysts.

What makes silicones flexible

Silicon and oxygen together simply make a material that is hard and brittle, just like quartz. But by combining the silica with organic materials, the result is a flexible compound that has all the advantages of both silica (it will stand up to high and low temperatures, is not affected by the weather or ultraviolet light, does not burn) and rubber (it is flexible, can be moulded, will set into a shape, is waterproof).

Silicones also have one extra important property: although they repel liquid water, they do allow water vapour to pass through (they are not gastight). This property allows waterproof clothing to "breathe". People can thus remain comfortable and dry inside waterproof clothing. Similarly, silicones painted on to timber or brick will stay watertight, but they can still breathe.

The only disadvantage of silicones is that they are not quite as flexible or strong as other types of rubber.

Some of the silicone rubbers used in the home (for example bath and sink sealants) can be squeezed out of tubes. They then set over the next few hours, sticking to the surfaces they have been applied to and curing, which emits a distinctive acrid smell (this is acetic acid gas). Other types of silicone actually cure using moisture from the air.

▲ Silicone sealant applied to a joint acts both as an adhesive and as water protection.

Silicon chips

A silicon chip is a piece of pure silicon that has been treated with other chemicals in a way that gives it special electrical properties.

A piece of silicon is normally a very good insulator, that is, it has a very high resistance to the flow of an electric current.

Materials only conduct electricity when they have mobile electrons that can flow when an electrical voltage is applied. Metals, for example, have many mobile electrons in their structure and so are very good conductors (they have a very low resistance).

In a silicon atom there are no electrons that are able to move and produce an electric current. However, scientists have been able to give silicon a resistance somewhere between an insulator and a conductor. This resulting material is called a semiconductor.

Making semiconducting materials

Silicon can be made semiconducting by combining phosphorus with silicon. As crystals of silicon and phosphorus grow, they bond together to leave one electron spare. The spare electrons are free to move, and therefore to conduct electricity.

Silicon with added phosphorus has a surplus of electrons; it is called a negative type or n-type semiconductor. Silicon with added aluminium, on the other hand, leaves a deficit of electrons. This makes it a positive, or p-type material.

By sandwiching the two materials and applying a voltage across them that attracts electrons from the n-type to the p-type, a current can be made to flow.

If a voltage is applied in the other direction, however, electrons cannot move. This kind of semiconductor is called a junction diode.

A diode

The simplest silicon chip is called a diode. It is a small piece of silicon that has two regions doped with metal. This is normally achieved by doping the whole of the chip with one impurity and then doping a small section with another impurity.

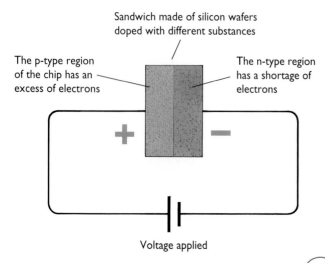

Sandwich made of silicon wafers doped with different substances

The p-type region of the chip has an excess of electrons

The n-type region has a shortage of electrons

+

−

Voltage applied

The uses of diodes

Junction diodes are widely used in electronics to detect signals in radios and act as switches in computers. But some diodes can also send out light when a voltage is applied across them. These are known as light emitting diodes, or LEDs. Light emitting diodes readily produce red, yellow and green light. They have an extremely long life and are now used as indictor lights for most electronics. The indicator light on the front of a computer monitor, for example, does not have a bulb behind it but an LED. And the colour of the light is not caused by a piece of green plastic; the colour you see is the colour of light emitted by the LED.

Some LEDs send out infra-red radiation. These can be used to send information along optical cables.

Crystals of ultra-pure silica

Ultra-pure silicon – needed for semiconductors – is obtained from silica by growing a large crystal from a tub of molten silica. This is done using the Czochralski method of crystal growth: A rod is pulled slowly out of the melt, leaving a growing crystal behind it.

diode: a semiconducting device that allows an electric current to flow in only one direction.

doping: the adding of metal atoms to a region of silicon to make it semiconducting.

electron: a tiny, negatively charged particle that is part of an atom. The flow of electrons through a solid material such as a wire produces an electric current.

semiconductor: a material of intermediate conductivity. Semiconductor devices often use silicon when they are made as part of diodes, transistors or integrated circuits.

Solar cells

The photoelectric effect is the release of electrons from semiconductors when light falls on their surfaces. Its use to generate electricity from sunlight has given some semiconductors the nickname "solar cells".

When light falls on the *junction* of a junction diode semiconductor, photons are absorbed, providing enough energy to allow some electrons to move across the junction. This produces an electric current; light is converted directly into electricity.

The solar cell junction diode is made so that as much light as possible falls on the junction. The surface layer of the diode sandwich is therefore made very thin and transparent.

A solar cell produces between 0.6 and 1.0 volts and a tiny current (a few milliAmps). To make the cells more powerful and give a better working voltage, cells are connected together.

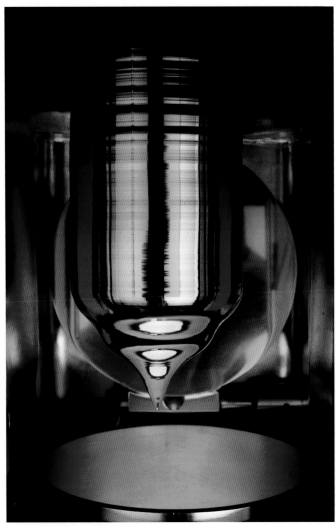

▲ A pure silicon crystal, which will be sliced up to make "chips".

▶ Solar cell arrays have to be placed in large banks if they are to generate substantial electricity. They also have to be placed in parts of the world where sunshine is common and the sun shines from as close to overhead as possible. Sites on or near the equator are ideal.

◄ Light emitting diodes can be used for low-energy consumption rear view lights designed to be fitted to bicycles.

Silicon and integrated circuits

Silicon has always been an important element in our lives through the silicates that surround us. But through the development of transistors and integrated circuits, this element has, in recent years, helped to transform the ways we do things, perhaps more than any other element on Earth. Here are some of the components that have made this revolution possible.

Transistors

The transistor lies at the heart of the electronics revolution. It is the fundamental building block of all computers and other solid-state circuits. Its invention in 1948 allowed large, cumbersome, unreliable and energy-consuming valves (vacuum tubes) to be replaced with small, robust, reliable devices that used very little energy.

Transistors, developed from junction diodes, have two junctions, which can be arranged to give two quite different effects. In one type the junctions are arranged back to back. This type has a single chip of silicon doped with one metal in the middle and a different metal at both ends. The central region of the transistor is called the base and it controls the flow of current through the chip from one end (called the emitter) to the other end (called the collector).

This kind of transistor is often used as an amplifier. A tiny current generated by, say, a voice from a microphone, arrives at the base of the transistor. A much larger current is flowing through from the emitter to the collector. The small base current interferes with this flow, making changes in the large current that mirror the changes in the small current. If this larger current is fed into a loudspeaker, an amplified sound is heard.

Alternatively, the same transistor can also be made to act as a very fast switch. This is how it is used in computers.

Since the first transistors were developed there have been many advances in how they are made. In particular, special types of transistor (known as MOSFETS or metal oxide semiconductor field effect transistors,) are simpler and need fewer components to complete their circuits. These are used as the main elements in microprocessors, where they are grouped by their thousands to make an integrated circuit or IC, commonly known as a "chip".

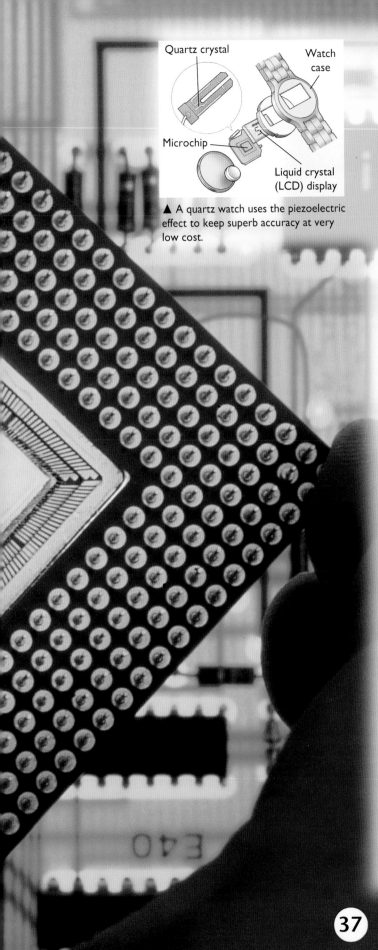

Quartz crystal

Watch case

Microchip

Liquid crystal (LCD) display

▲ A quartz watch uses the piezoelectric effect to keep superb accuracy at very low cost.

The crystal structure of quartz

The crystal structure of quartz is responsible for many of its unique properties. For example, a quartz crystal placed under pressure will produce electrical charges at each end of the crystal. A voltage applied across it results in minute changes in shape. These two phenomena make up the piezoelectric effect.

The piezoelectric effect can be used as a pressure gauge or an electrical oscillator (a device that vibrates). Quartz crystal oscillators are at the heart of every digital watch and the clocks that help to control computers.

Quartz oscillators are also used in radios and other receiving and transmitting devices.

ICs (integrated circuits)

By carefully placing the doping impurities on a wafer of silicon, the wafer can be made to behave as though it were a myriad of separate components.

This is possible because silicon can be made to change its conductivity over very short distances. In turn, this makes it possible for the circuit components to be microscopically small.

The principle of an IC is simple, but making ICs is extremely difficult because of the miniaturisation and accuracy required. A circuit is first drawn at a large-scale and made up as a photographic transparency. A light is shone through the transparency and focused onto a silver-based, light-sensitive coating on the surface of the chip.

The chip surface is next etched so that only the silver that will make the circuit remains. At the same time impurities have to be introduced into the surface of the wafer so that the chip can act as transistors, diodes, resistors, etc., and thus complete the circuit.

The whole of this circuit, and the aluminium leads that will be used to connect the IC to the "wider world" are encased in a plastic block.

◀ A modern IC, showing the multitude of external connections.

Glass

When silica is heated to very high temperatures (about 1600°C), the natural bonds of the silica break down and the crystals change to a noncrystalline amorphous, glassy material called fused silica. If metal compounds are added to the melt, the result is the transparent material we know as glass.

All glass contains a natural glass, for example silica in the form of sand and a flux such as sodium carbonate to make the glass melt more easily. Glass and a flux alone are not stable and, for example, such glass is soluble in water (see water glass on page 6). As a result, a stabiliser, in the form of calcium carbonate (limestone) must be added.

The sodium–calcium–silicon glass is called soda lime glass. About nine-tenths of all glass is soda lime glass. Its main use is as windows and bottles. It has about one-eighth soda, one-eighth lime and three-quarters sand (silica).

Sometimes boron oxide is added to the melt. This changes the property of the glass so that it does not expand or contract significantly with changes in temperature. This means the glass will not crack when heated or cooled suddenly. This is the form of glass used in most cookware applications.

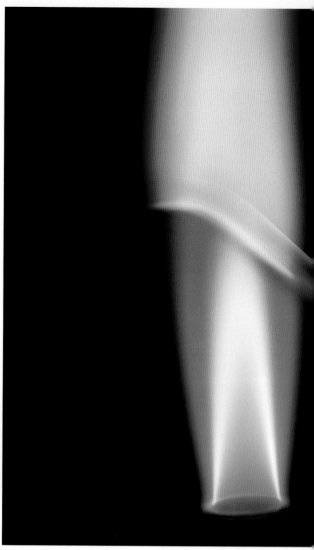

▲ Silica, soda and lime react to form soda lime glass. The addition of these materials splits up the tight structure of pure quartz, lowering its melting point. This makes it less viscous and easier to melt and shape.

The top picture shows a rod of soda lime glass bending in a Bunsen flame. Notice the flame colour has been changed because of the presence of sodium and calcium atoms, which produce orange and red colours, respectively.

The structure of glass

Glass is noncrystalline, or amorphous. The silicon atom is completely surrounded by four oxygen atoms. Each silica molecule touches those adjacent to it and shares oxygen atoms, making randomly arranged chains and rings. The sodium and calcium ions in soda lime glass fit in between the rings and help to lock them together.

EQUATION: Chemical change in soda glass

Sodium carbonate + silica ⇨ soda glass + carbon dioxide

$$Na_2CO_3 + SiO_2(s) \Rightarrow Na_2SiO_3(s) + CO_2(g)$$

EQUATION: Chemical change in lime glass

Calcium carbonate + silica ⇨ lime glass + carbon dioxide

$$CaCO_3(s) + SiO_2(s) \Rightarrow CaSiO_3(s) + CO_2(g)$$

▼ Glass cracks because it is a poor conductor of heat. As a result, when one side is heated, the other side remains cool. As the heated side tries to expand, the glass cracks. Borosilicate glass not only has a high melting point, but it resists changes in size both on heating and cooling and thus prevents cracking. In this borosilicate tube, sodium chloride has been melted. The melting point of sodium chloride (common salt) is **809°C**.

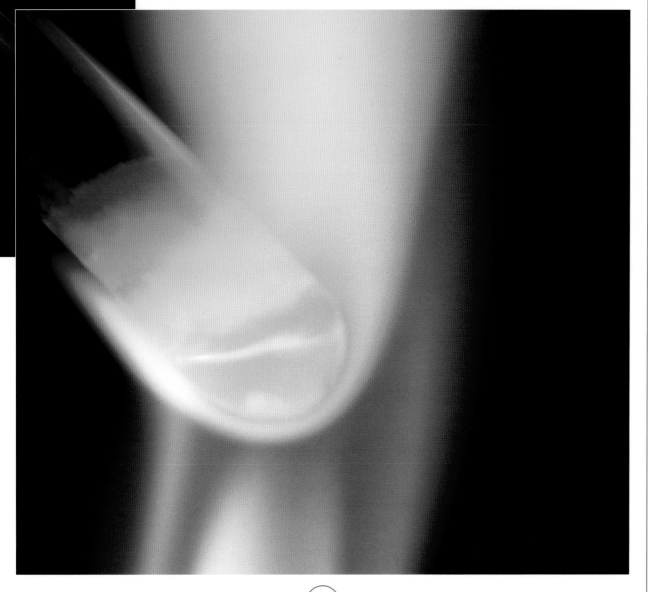

Making glass

Float glass is the name given to large flat sheets of glass that are made by floating molten glass on molten tin.

The manufacture of float glass relies on several unique properties of tin. Tin and glass do not react; tin is denser than glass; and whereas the melting point of tin (232°C) is far below the melting point of glass (1400°C) the boiling point of tin is well above the melting point of glass. Thus tin can be used as a stable, nonreacting liquid on which to float the glass.

During the continuous formation of flat glass, molten glass is run out over a very shallow bed of molten tin, ensuring that the glass cools as a smooth, flat sheet.

▲ The traditional method of blowing glass. The almost molten glass is collected on the end of a pipe and a skilled worker blows down the tube and into the molten ball.

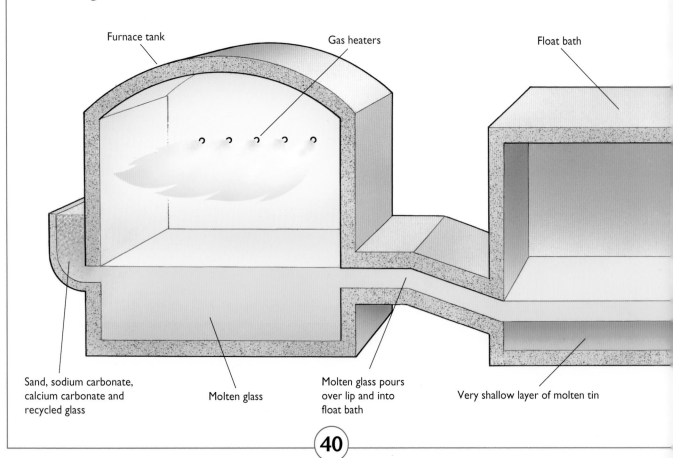

Furnace tank

Gas heaters

Float bath

Sand, sodium carbonate, calcium carbonate and recycled glass

Molten glass

Molten glass pours over lip and into float bath

Very shallow layer of molten tin

Coloured glass

Coloured glass is made by adding a variety of metal compounds to the melt, usually in quantities that comprise less than half of 1% of the glass. Soda lime glass is normally a very pale green, mainly because it contains impurities of iron. But to get a richer colour more impurities must be added. A rich red is obtained by adding cadmium sulphide and selenide. Cobalt is added to produce a deep blue glass. Brown is obtained by adding iron sulphide. It is used for some beer bottles. Manganese gives pink and violet colours.

▼ This picture is a piece of Venetian glassware, which has been skilfully moulded at a temperature near the melting point of glass.

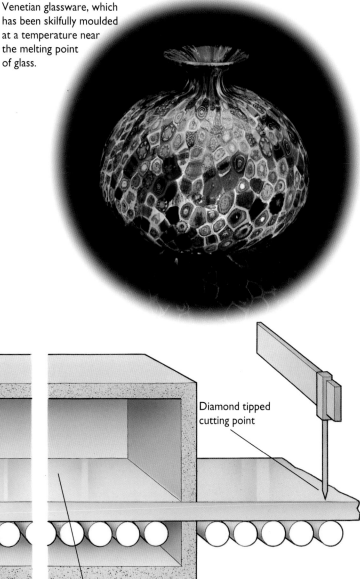

Glass sheet

Controlled atmosphere

Diamond tipped cutting point

Roller

The glass is slowly cooled in a long chamber called the annealing lehr to allow the release of internal stresses in the glass

41

Using silica to predict eruptions

All lavas are silicates. They are the source of the world's sandstones, clays and other rocks and materials. But volcanic eruptions can produce great danger for people living quite close to them, and any way of predicting the nature of the next explosion will help to save lives. One of the most promising developments of recent years has been in the investigation of the chemistry of lavas, especially the amount of silica they contain.

▼ A piece of acid lava from a volcano. This material is high in silica.

◀ Basaltic material oozing out from a Hawaiian volcano. Basalt is very low in silicon and so will not produce violent eruptions.

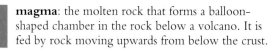

magma: the molten rock that forms a balloon-shaped chamber in the rock below a volcano. It is fed by rock moving upwards from below the crust.

viscous: slow moving, syrupy. A liquid that has a low viscosity is said to be mobile.

Chemistry of magma chambers

The minerals that pour over the Earth's surface or that are sent high into the sky as ash, are formed at great depth in pockets of molten rock known as magma chambers.

Magma chambers do not have a uniform chemical composition. This is because they receive new materials from below and, from time to time, they send newly formed minerals to the surface.

Scientists have found that the type of minerals created depend on the length of time the magma sits in the magma chamber before it is emptied and refreshed from below.

The longer the waiting period (the longer the volcano is dormant), the more the minerals become dominated by silica as opposed to the other mineral-forming elements. What happens is a kind of separating process, where the silica rises to the top of the chamber, and the other heavier elements are pushed below.

An eruption of rock rich in silica (say between 65 and 75% silica) does not flow easily. It is called a viscous lava. As a result it quickly solidifies as it reaches the air. The pressure of the rising magma behind, causes this syrupy material to be blown to pieces, leading to an explosion of great violence and the ejection of huge volumes of fragments of lava, which cool to form ash as they fall through the air. On the other hand, if the magma has been refreshed relatively recently, the magma has not had time to separate out and it remains less rich in silica (say 45–55%). As a result the lava is runny and able to flow from the volcano easily, with little build up of pressure. It is rich in dark iron and manganese minerals such as olivine and hornblende. An eruption of this lava is not as violent and therefore less liable to lead to disaster.

Scientists can look at the materials surrounding a volcano and find out how old they are. They can also examine the minerals they contain. From this information, it is possible to work out how often the volcano erupts on average, and what kind of material it ejects on each occasion. Using this chemical information, and the length of time since the last eruption, the degree of violence of the next eruption can be predicted.

◀ The violent eruption of Mt St Helens in 1980 was due to the high silica content of the magma in the chamber below the volcano.

Key facts about...

Silicon

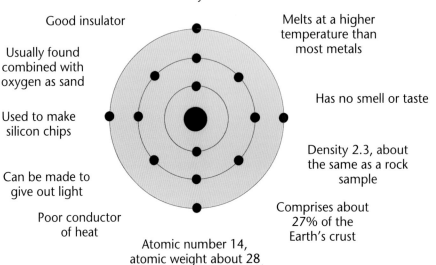

A grey element, chemical symbol Si

Good insulator

Usually found combined with oxygen as sand

Used to make silicon chips

Can be made to give out light

Poor conductor of heat

Melts at a higher temperature than most metals

Has no smell or taste

Density 2.3, about the same as a rock sample

Comprises about 27% of the Earth's crust

Atomic number 14, atomic weight about 28

SHELL DIAGRAMS

The shell diagram on this page represents an atom of the element silicon. The total number of electrons is shown in the relevant orbitals, or shells, around the central nucleus.

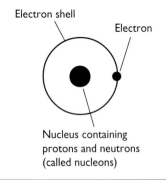

Electron shell

Electron

Nucleus containing protons and neutrons (called nucleons)

▶▶ The small picture shows a ceramic crucible on a pipe triangle being used in a high-temperature laboratory experiment. The large picture is a fine example of hand blown glass. The dolphin is inside the bottle.

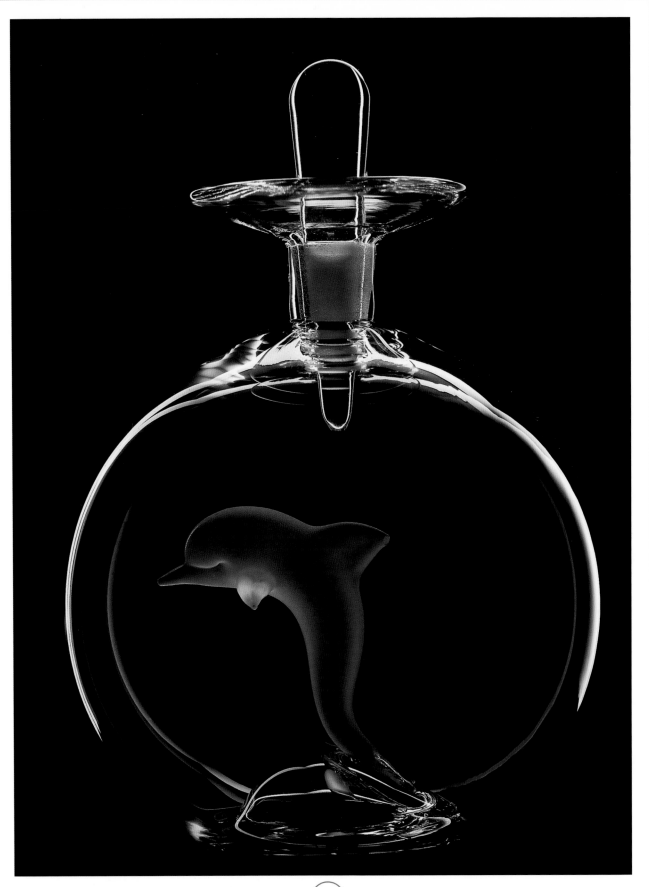

The Periodic Table

The Periodic Table sets out the relationships among the elements of the Universe. According to the Periodic Table, certain elements fall into groups. The pattern of these groups has, in the past, allowed scientists to predict elements that had not at that time been discovered. It can still be used today to predict the properties of unfamiliar elements.

The Periodic Table was first described by a Russian teacher, Dmitry Ivanovich Mendeleev, between 1869 and 1870. He was interested in writing a chemistry textbook, and wanted to show his students that there were certain patterns in the elements that had been discovered. So he set out the elements (of which there were 57 at the time) according to their known properties. On the assumption that there was pattern to the elements, he left blank spaces where elements seemed to be missing. Using this first version of the Periodic Table, he was able to predict in detail the chemical and physical properties of elements that had not yet been discovered. Other scientists began to look for the missing elements, and they soon found them.

GROUP

Legend:
- Metals
- Metalloids (semi-metals)
- Non-metals

Transition metals

1	2						
1 **H** Hydrogen 1							
3 **Li** Lithium 7	4 **Be** Beryllium 9						
11 **Na** Sodium 23	12 **Mg** Magnesium 24						
19 **K** Potassium 39	20 **Ca** Calcium 40	21 **Sc** Scandium 45	22 **Ti** Titanium 48	23 **V** Vanadium 51	24 **Cr** Chromium 52	25 **Mn** Manganese 55	26 **Fe** Iron 56
37 **Rb** Rubidium 85	38 **Sr** Strontium 88	39 **Y** Yttrium 89	40 **Zr** Zirconium 91	41 **Nb** Niobium 93	42 **Mo** Molybdenum 96	43 **Tc** Technetium (99)	44 **Ru** Ruthenium 101
55 **Cs** Cesium 133	56 **Ba** Barium 137	71 **Lu** Lutetium 175	72 **Hf** Hafnium 178	73 **Ta** Tantalum 181	74 **W** Tungsten 184	75 **Re** Rhenium 186	76 **Os** Osmium 190
87 **Fr** Francium 223	88 **Ra** Radium 226	103 **Lw** Lawrencium (260)	104 **Unq** Unnilquadium (261)	105 **Unp** Unnilpentium (262)	106 **Unh** Unnilhexium (263)	107 **Uns** Unnilseptium (262)	108 **Uno** Unniloctium (265)

Lanthanide metals

57 **La** Lanthanum 139	58 **Ce** Cerium 140	59 **Pr** Praseodymium 141	60 **Nd** Neodymium 144

Actinoid metals

89 **Ac** Actinium (227)	90 **Th** Thorium 232	91 **Pa** Protactinium 231	92 **U** Uranium 238

Hydrogen did not seem to fit into the table, so he placed it in a box on its own. Otherwise the elements were all placed horizontally. When an element was reached with properties similar to the first one in the top row, a second row was started. By following this rule, similarities among the elements can be found by reading up and down. By reading across the rows, the elements progressively increase their atomic number. This number indicates the number of positively charged particles (protons) in the nucleus of each atom. This is also the number of negatively charged particles (electrons) in the atom.

The chemical properties of an element depend on the number of electrons in the outermost shell.

Atoms can form compounds by sharing electrons in their outermost shells. This explains why atoms with a full set of electrons (like helium, an inert gas) are unreactive, whereas atoms with an incomplete electron shell (such as chlorine) are very reactive. Elements can also combine by the complete transfer of electrons from metals to non-metals and the compounds formed contain ions.

Radioactive elements lose particles from their nucleus and electrons from their surrounding shells. As a result their atomic number changes and they become new elements.

Key:

Atomic (proton) number	13
Symbol	Al
Name	Aluminium
Approximate relative atomic mass (Approximate atomic weight)	27

3	4	5	6	7	0
					2 **He** Helium 4
5 **B** Boron 11	6 **C** Carbon 12	7 **N** Nitrogen 14	8 **O** Oxygen 16	9 **F** Fluorine 19	10 **Ne** Neon 20
13 **Al** Aluminium 27	14 **Si** Silicon 28	15 **P** Phosphorus 31	16 **S** Sulphur 32	17 **Cl** Chlorine 35	18 **Ar** Argon 40

27 **Co** Cobalt 59	28 **Ni** Nickel 59	29 **Cu** Copper 64	30 **Zn** Zinc 65	31 **Ga** Gallium 70	32 **Ge** Germanium 73	33 **As** Arsenic 75	34 **Se** Selenium 79	35 **Br** Bromine 80	36 **Kr** Krypton 84
45 **Rh** Rhodium 103	46 **Pd** Palladium 106	47 **Ag** Silver 108	48 **Cd** Cadmium 112	49 **In** Indium 115	50 **Sn** Tin 119	51 **Sb** Antimony 122	52 **Te** Tellurium 128	53 **I** Iodine 127	54 **Xe** Xenon 131
77 **Ir** Iridium 192	78 **Pt** Platinum 195	79 **Au** Gold 197	80 **Hg** Mercury 201	81 **Tl** Thallium 204	82 **Pb** Lead 207	83 **Bi** Bismuth 209	84 **Po** Polonium (209)	85 **At** Astatine (210)	86 **Rn** Radon (222)

109 **Une** Unnilennium (266)

61 **Pm** Promethium (145)	62 **Sm** Samarium 150	63 **Eu** Europium 152	64 **Gd** Gadolinium 157	65 **Tb** Terbium 159	66 **Dy** Dysprosium 163	67 **Ho** Holmium 165	68 **Er** Erbium 167	69 **Tm** Thulium 169	70 **Yb** Ytterbium 173
93 **Np** Neptunium (237)	94 **Pu** Plutonium (244)	95 **Am** Americium (243)	96 **Cm** Curium (247)	97 **Bk** Berkelium (247)	98 **Cf** Californium (251)	99 **Es** Einsteinium (252)	100 **Fm** Fermium (257)	101 **Md** Mendelevium (258)	102 **No** Nobelium (259)

Understanding equations

As you read through this book, you will notice that many pages contain equations using symbols. If you are not familiar with these symbols, read this page. Symbols make it easy for chemists to write out the reactions that are occurring in a way that allows a better understanding of the processes involved.

Symbols for the elements

The basis of the modern use of symbols for elements dates back to the 19th century. At this time a shorthand was developed using the first letter of the element wherever possible. Thus "O" stands for oxygen, "H" stands for hydrogen

and so on. However, if we were to use only the first letter, then there could be some confusion. For example, nitrogen and nickel would both use the symbols N. To overcome this problem, many elements are symbolised using the first two letters of their full name, and the second letter is lowercase. Thus although nitrogen is N, nickel becomes Ni. Not all symbols come from the English name; many use the Latin name instead. This is why, for example, gold is not G but Au (for the Latin *aurum*) and sodium has the symbol Na, from the Latin *natrium*.

Compounds of elements are made by combining letters. Thus the molecule carbon

Written and symbolic equations

In this book, important chemical equations are briefly stated in words (these are called word equations), and are then shown in their symbolic form along with the states.

What reaction the equation illustrates

Word equation

Symbol equation

Sometimes you will find additional descriptions below the symbolic equation.

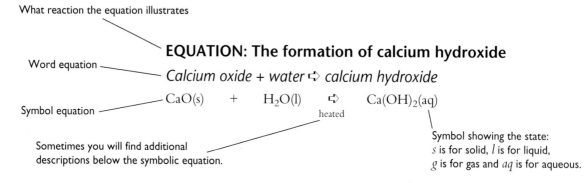

EQUATION: The formation of calcium hydroxide

Calcium oxide + water ⇨ calcium hydroxide

$$CaO(s) \quad + \quad H_2O(l) \quad ⇨ \quad Ca(OH)_2(aq)$$

heated

Symbol showing the state:
s is for solid, l is for liquid,
g is for gas and aq is for aqueous.

Diagrams

Some of the equations are shown as graphic representations.

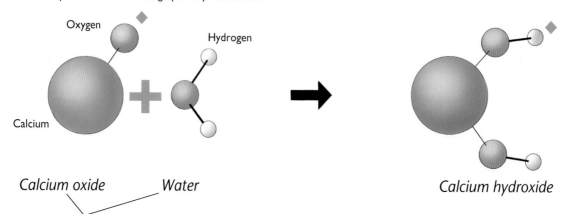

Oxygen

Hydrogen

Calcium

Calcium oxide   Water

Calcium hydroxide

Sometimes the written equation is broken up and put below the relevant stages in the graphic representation.

monoxide is CO. By using lowercase letters for the second letter of an element, it is possible to show that cobalt, symbol Co, is not the same as the molecule carbon monoxide, CO.

However, the letters can be made to do much more than this. In many molecules, atoms combine in unequal numbers. So, for example, carbon dioxide has one atom of carbon for every two of oxygen. This is shown by using the number 2 beside the oxygen, and the symbol becomes CO_2.

In practice, some groups of atoms combine as a unit with other substances. Thus, for example, calcium bicarbonate (one of the compounds used in some antacid pills) is written $Ca(HCO_3)_2$. This shows that the part of the substance inside the brackets reacts as a unit and the "2" outside the brackets shows the presence of two such units.

Some substances attract water molecules to themselves. To show this a dot is used. Thus the blue form of copper sulphate is written $CuSO_4.5H_2O$. In this case five molecules of water attract to one of copper sulphate.

When you see the dot, you know that this water can be driven off by heating; it is part of the crystal structure.

In a reaction substances change by rearranging the combinations of atoms. The way they change is shown by using the chemical symbols, placing those that will react (the starting materials, or reactants) on the left and the products of the reaction on the right. Between the two, chemists use an arrow to show which way the reaction is occurring.

It is possible to describe a reaction in words. This gives word equations, which are given throughout this book. However, it is easier to understand what is happening by using an equation containing symbols. These are also given in many places. They are not given when the equations are very complex.

In any equation both sides balance; that is, there must be an equal number of like atoms on both sides of the arrow. When you try to write down reactions, you, too, must balance your equation; you cannot have a few atoms left over at the end!

The symbols in brackets are abbreviations for the physical state of each substance taking part, so that (s) is used for solid, (l) for liquid, (g) for gas and (aq) for an aqueous solution, that is, a solution of a substance dissolved in water.

Atoms and ions
Each sphere represents a particle of an element. A particle can be an atom or an ion. Each atom or ion is associated with other atoms or ions through bonds – forces of attraction. The size of the particles and the nature of the bonds can be extremely important in determining the nature of the reaction or the properties of the compound.

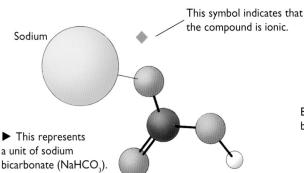

Sodium

This symbol indicates that the compound is ionic.

▶ This represents a unit of sodium bicarbonate ($NaHCO_3$).

The term "unit" is sometimes used to simplify the representation of a combination of ions.

Chemical symbols, equations and diagrams
The arrangement of any molecule or compound can be shown in one of the two ways shown below, depending on which gives the clearer picture. The left-hand diagram is called a ball-and-stick diagram because it uses rods and spheres to show the structure of the material. This example shows water, H_2O. There are two hydrogen atoms and one oxygen atom.

Bond shown by "stick"

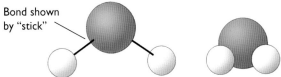

Colours too
The colours of each of the particles help differentiate the elements involved. The diagram can then be matched to the written and symbolic equation given with the diagram. In the case above, oxygen is red and hydrogen is grey.

Glossary of technical terms

absorb: to soak up a substance. Compare to adsorb.

acetone: a petroleum-based solvent.

acid: compounds containing hydrogen which can attack and dissolve many substances. Acids are described as weak or strong, dilute or concentrated, mineral or organic.

acidity: a general term for the strength of an acid in a solution.

acid rain: rain that is contaminated by acid gases such as sulphur dioxide and nitrogen oxides released by pollution.

adsorb/adsorption: to "collect" gas molecules or other particles on to the *surface* of a substance. They are not chemically combined and can be removed. (The process is called "adsorption".) Compare to absorb.

alchemy: the traditional "art" of working with chemicals that prevailed through the Middle Ages. One of the main challenges of alchemy was to make gold from lead. Alchemy faded away as scientific chemistry was developed in the 17th century.

alkali: a base in solution.

alkaline: the opposite of acidic. Alkalis are bases that dissolve, and alkaline materials are called basic materials. Solutions of alkalis have a pH greater than 7.0 because they contain relatively few hydrogen ions.

alloy: a mixture of a metal and various other elements.

alpha particle: a stable combination of two protons and two neutrons, which is ejected from the nucleus of a radioactive atom as it decays. An alpha particle is also the nucleus of the atom of helium. If it captures two electrons it can become a neutral helium atom.

amalgam: a liquid alloy of mercury with another metal.

amino acid: amino acids are organic compounds that are the building blocks for the proteins in the body.

amorphous: a solid in which the atoms are not arranged regularly (i.e. "glassy"). Compare with crystalline.

amphoteric: a metal that will react with both acids and alkalis.

anhydrous: a substance from which water has been removed by heating. Many hydrated salts are crystalline. When they are heated and the water is driven off, the material changes to an anhydrous powder.

anion: a negatively charged atom or group of atoms.

anode: the negative terminal of a battery or the positive electrode of an electrolysis cell.

anodising: a process that uses the effect of electrolysis to make a surface corrosion-resistant.

antacid: a common name for any compound that reacts with stomach acid to neutralise it.

antioxidant: a substance that prevents oxidation of some other substance.

aqueous: a solid dissolved in water. Usually used as "aqueous solution".

atom: the smallest particle of an element.

atomic number: the number of electrons or the number of protons in an atom.

atomised: broken up into a very fine mist. The term is used in connection with sprays and engine fuel systems.

aurora: the "northern lights" and "southern lights" that show as coloured bands of light in the night sky at high latitudes. They are associated with the way cosmic rays interact with oxygen and nitrogen in the air.

basalt: an igneous rock with a low proportion of silica (usually below 55%). It has microscopically small crystals.

base: a compound that may be soapy to the touch and that can react with an acid in water to form a salt and water.

battery: a series of electrochemical cells.

bauxite: an ore of aluminium, of which about half is aluminium oxide.

becquerel: a unit of radiation equal to one nuclear disintegration per second.

beta particle: a form of radiation in which electrons are emitted from an atom as the nucleus breaks down.

bleach: a substance that removes stains from materials either by oxidising or reducing the staining compound.

boiling point: the temperature at which a liquid boils, changing from a liquid to a gas.

bond: chemical bonding is either a transfer or sharing of electrons by two or more atoms. There are a number of types of chemical bond, some very strong (such as covalent bonds), others weak (such as hydrogen bonds). Chemical bonds form because the linked molecule is more stable than the unlinked atoms from which it formed. For example, the hydrogen molecule (H_2) is more stable than single atoms of hydrogen, which is why hydrogen gas is always found as molecules of two hydrogen atoms.

brass: a metal alloy principally of copper and zinc.

brazing: a form of soldering, in which brass is used as the joining metal.

brine: a solution of salt (sodium chloride) in water.

bronze: an alloy principally of copper and tin.

buffer: a chemistry term meaning a mixture of substances in solution that resists a change in the acidity or alkalinity of the solution.

capillary action: the tendency of a liquid to be sucked into small spaces, such as between objects and through narrow-pore tubes. The force to do this comes from surface tension.

catalyst: a substance that speeds up a chemical reaction but itself remains unaltered at the end of the reaction.

cathode: the positive terminal of a battery or the negative electrode of an electrolysis cell.

cathodic protection: the technique of making the object that is to be protected from corrosion into the cathode of a cell. For example, a material, such as steel, is protected by coupling it with a more reactive metal, such as magnesium. Steel forms the cathode and magnesium the anode. Zinc protects steel in the same way.

cation: a positively charged atom or group of atoms.

caustic: a substance that can cause burns if it touches the skin.

cell: a vessel containing two electrodes and an electrolyte that can act as an electrical conductor.

ceramic: a material based on clay minerals, which has been heated so that it has chemically hardened.

chalk: a pure form of calcium carbonate made of the crushed bodies of microscopic sea creatures, such as plankton and algae.

change of state: a change between one of the three states of matter, solid, liquid and gas.

chlorination: adding chlorine to a substance.

cladding: a surface sheet of material designed to protect other materials from corrosion.

clay: a microscopically small plate-like mineral that makes up the bulk of many soils. It has a sticky feel when wet.

combustion: the special case of oxidisation of a substance where a considerable amount of heat and usually light are given out. Combustion is often referred to as "burning".

compound: a chemical consisting of two or more elements chemically bonded together. Calcium atoms can combine with carbon atoms and oxygen atoms to make calcium carbonate, a compound of all three atoms.

condensation nuclei: microscopic particles of dust, salt and other materials suspended in the air, which attract water molecules.

conduction: (i) the exchange of heat (heat conduction) by contact with another object or (ii) allowing the flow of electrons (electrical conduction).

convection: the exchange of heat energy with the surroundings produced by the flow of a fluid due to being heated or cooled.

corrosion: the *slow* decay of a substance resulting from contact with gases and liquids in the environment. The term is often applied to metals. Rust is the corrosion of iron.

corrosive: a substance, either an acid or an alkali, that *rapidly* attacks a wide range of other substances.

cosmic rays: particles that fly through space and bombard all atoms on the Earth's surface. When they interact with the atmosphere they produce showers of secondary particles.

covalent bond: the most common form of strong chemical bonding, which occurs when two atoms *share* electrons.

cracking: breaking down complex molecules into simpler components. It is a term particularly used in oil refining.

crude oil: a chemical mixture of petroleum liquids. Crude oil forms the raw material for an oil refinery.

crystal: a substance that has grown freely so that it can develop external faces. Compare with crystalline, where the atoms are not free to form individual crystals and amorphous where the atoms are arranged irregularly.

crystalline: the organisation of atoms into a rigid "honeycomb-like" pattern without distinct crystal faces.

crystal systems: seven patterns or systems into which all of the world's crystals can be grouped. They are: cubic, hexagonal, rhombohedral, tetragonal, orthorhombic, monoclinic and triclinic.

cubic crystal system: groupings of crystals that look like cubes.

curie: a unit of radiation. The amount of radiation emitted by 1 g of radium each second. (The curie is equal to 37 billion becquerels.)

current: an electric current is produced by a flow of electrons through a conducting solid or ions through a conducting liquid.

decay (radioactive decay): the way that a radioactive element changes into another element because of loss of mass through radiation. For example uranium decays (changes) to lead.

decompose: to break down a substance (for example by heat or with the aid of a catalyst) into simpler components. In such a chemical reaction only one substance is involved.

dehydration: the removal of water from a substance by heating it, placing it in a dry atmosphere, or through the action of a drying agent.

density: the mass per unit volume (e.g. g/cc).

desertification: a process whereby a soil is allowed to become degraded to a state in which crops can no longer grow, i.e. desert-like. Chemical desertification is usually the result of contamination with halides because of poor irrigation practices.

detergent: a petroleum-based chemical that removes dirt.

diaphragm: a semipermeable membrane – a kind of ultra-fine mesh filter – that will allow only small ions to pass through. It is used in the electrolysis of brine.

diffusion: the slow mixing of one substance with another until the two substances are evenly mixed.

digestive tract: the system of the body that forms the pathway for food and its waste products. It begins at the mouth and includes the stomach and the intestines.

dilute acid: an acid whose concentration has been reduced by a large proportion of water.

diode: a semiconducting device that allows an electric current to flow in only one direction.

disinfectant: a chemical that kills bacteria and other microorganisms.

dissociate: to break apart. In the case of acids it means to break up forming hydrogen ions. This is an example of ionisation. Strong acids dissociate completely. Weak acids are not completely ionised and a solution of a weak acid has a relatively low concentration of hydrogen ions.

dissolve: to break down a substance in a solution without a resultant reaction.

distillation: the process of separating mixtures by condensing the vapours through cooling.

doping: adding metal atoms to a region of silicon to make it semiconducting.

dye: a coloured substance that will stick to another substance, so that both appear coloured.

electrode: a conductor that forms one terminal of a cell.

electrolysis: an electrical–chemical process that uses an electric current to cause the break up of a compound and the movement of metal ions in a solution. The process happens in many natural situations (as for example in rusting) and is also commonly used in industry for purifying (refining) metals or for plating metal objects with a fine, even metal coating.

electrolyte: a solution that conducts electricity.

electron: a tiny, negatively charged particle that is part of an atom. The flow of electrons through a solid material such as a wire produces an electric current.

electroplating: depositing a thin layer of a metal onto the surface of another substance using electrolysis.

element: a substance that cannot be decomposed into simpler substances by chemical means

emulsion: tiny droplets of one substance dispersed in another. A common oil in water emulsion is milk. The tiny droplets in an emulsion tend to come together, so another stabilising substance is often needed to wrap the particles of grease and oil in a stable coat. Soaps and detergents are such agents. Photographic film is an example of a solid emulsion.

endothermic reaction: a reaction that takes heat from the surroundings. The reaction of carbon monoxide with a metal oxide is an example.

enzyme: organic catalysts in the form of proteins in the body that speed up chemical reactions. Every living cell contains hundreds of enzymes, which ensure that the processes of life continue. Should enzymes be made inoperative, such as through mercury poisoning, then death follows.

ester: organic compounds, formed by the reaction of an alcohol with an acid, which often have a fruity taste.

evaporation: the change of state of a liquid to a gas. Evaporation happens below the boiling point and is used as a method of separating out the materials in a solution.

exothermic reaction: a reaction that gives heat to the surroundings. Many oxidation reactions, for example, give out heat.

explosive: a substance which, when a shock is applied to it, decomposes very rapidly, releasing a very large amount of heat and creating a large volume of gases as a shock wave.

extrusion: forming a shape by pushing it through a die. For example, toothpaste is extruded through the cap (die) of the toothpaste tube.

fallout: radioactive particles that reach the ground from radioactive materials in the atmosphere.

fat: semi-solid energy-rich compounds derived from plants or animals and which are made of carbon, hydrogen and oxygen. Scientists call these esters.

feldspar: a mineral consisting of sheets of aluminium silicate. This is the mineral from which the clay in soils is made.

fertile: able to provide the nutrients needed for unrestricted plant growth.

filtration: the separation of a liquid from a solid using a membrane with small holes.

fission: the breakdown of the structure of an atom, popularly called "splitting the atom" because the atom is split into approximately two other nuclei. This is different from, for example, the small change that happens when radioactivity is emitted.

fixation of nitrogen: the processes that natural organisms, such as bacteria, use to turn the nitrogen of the air into ammonium compounds.

fixing: making solid and liquid nitrogen-containing compounds from nitrogen gas. The compounds that are formed can be used as fertilisers.

fluid: able to flow; either a liquid or a gas.

fluorescent: a substance that gives out visible light when struck by invisible waves such as ultraviolet rays.

flux: a material used to make it easier for a liquid to flow. A flux dissolves metal oxides and so prevents a metal from oxidising while being heated.

foam: a substance that is sufficiently gelatinous to be able to contain bubbles of gas. The gas bulks up the substance, making it behave as though it were semi-rigid.

fossil fuels: hydrocarbon compounds that have been formed from buried plant and animal remains. High pressures and temperatures lasting over millions of years are required. The fossil fuels are coal, oil and natural gas.

fraction: a group of similar components of a mixture. In the petroleum industry the light fractions of crude oil are those with the smallest molecules, while the medium and heavy fractions have larger molecules.

free radical: a very reactive atom or group with a "spare" electron.

freezing point: the temperature at which a substance changes from a liquid to a solid. It is the same temperature as the melting point.

fuel: a concentrated form of chemical energy. The main sources of fuels (called fossil fuels because they were formed by geological processes) are coal, crude oil and natural gas. Products include methane, propane and gasoline. The fuel for stars and space vehicles is hydrogen.

fuel rods: rods of uranium or other radioactive material used as a fuel in nuclear power stations.

fuming: an unstable liquid that gives off a gas. Very concentrated acid solutions are often fuming solutions.

fungicide: any chemical that is designed to kill fungi and control the spread of fungal spores.

fusion: combining atoms to form a heavier atom.

galvanising: applying a thin zinc coating to protect another metal.

gamma rays: waves of radiation produced as the nucleus of a radioactive element rearranges itself into a tighter cluster of protons and neutrons. Gamma rays carry enough energy to damage living cells.

gangue: the unwanted material in an ore.

gas: a form of matter in which the molecules form no definite shape and are free to move about to fill any vessel they are put in.

gelatinous: a term meaning made with water. Because a gelatinous precipitate is mostly water, it is of a similar density to water and will float or lie suspended in the liquid.

gelling agent: a semi-solid jelly-like substance.

gemstone: a wide range of minerals valued by people, both as crystals (such as emerald) and as decorative stones (such as agate). There is no single chemical formula for a gemstone.

glass: a transparent silicate without any crystal growth. It has a glassy lustre and breaks with a curved fracture. Note that some minerals have all these features and are therefore natural glasses. Household glass is a synthetic silicate.

glucose: the most common of the natural sugars. It occurs in the polymer known as cellulose, the fibre in plants. Starch is also a form of glucose. The breakdown of glucose provides the energy that animals need for life.

granite: an igneous rock with a high proportion of silica (usually over 65%). It has well-developed large crystals. The largest pink, grey or white crystals are feldspar.

Greenhouse Effect: an increase of the global air temperature as a result of heat released from burning fossil fuels being absorbed by carbon dioxide in the atmosphere.

gypsum: the name for calcium sulphate. It is commonly found as Plaster of Paris and wallboards.

half-life: the time it takes for the radiation coming from a sample of a radioactive element to decrease by half.

halide: a salt of one of the halogens (fluorine, chlorine, bromine and iodine).

halite: the mineral made of sodium chloride.

halogen: one of a group of elements including chlorine, bromine, iodine and fluorine.

heat-producing: see exothermic reaction.

high explosive: a form of explosive that will only work when it receives a shock from another explosive. High explosives are much more powerful than ordinary explosives. Gunpowder is not a high explosive.

hydrate: a solid compound in crystalline form that contains molecular water. Hydrates commonly form when a solution of a soluble salt is evaporated. The water that forms part of a hydrate crystal is known as the "water of crystallization". It can usually be removed by heating, leaving an anhydrous salt.

hydration: the absorption of water by a substance. Hydrated materials are not "wet" but remain firm, apparently dry, solids. In some cases, hydration makes the substance change colour, in many other cases there is no colour change, simply a change in volume.

hydrocarbon: a compound in which only hydrogen and carbon atoms are present. Most fuels are hydrocarbons, as is the simple plastic polyethene (known as polythene).

hydrogen bond: a type of attractive force that holds one molecule to another. It is one of the weaker forms of intermolecular attractive force.

hydrothermal: a process in which hot water is involved. It is usually used in the context of rock formation because hot water and other fluids sent outwards from liquid magmas are important carriers of metals and the minerals that form gemstones.

igneous rock: a rock that has solidified from molten rock, either volcanic lava on the Earth's surface or magma deep underground. In either case the rock develops a network of interlocking crystals.

incendiary: a substance designed to cause burning.

indicator: a substance or mixture of substances that change colour with acidity or alkalinity.

inert: nonreactive.

infra-red radiation: a form of light radiation where the wavelength of the waves is slightly longer than visible light. Most heat radiation is in the infra-red band.

insoluble: a substance that will not dissolve.

ion: an atom, or group of atoms, that has gained or lost one or more electrons and so developed an electrical charge. Ions behave differently from electrically neutral atoms and molecules. They can move in an electric field, and they can also bind strongly to solvent molecules such as water. Positively charged ions are called cations; negatively charged ions are called anions. Ions carry electrical current through solutions.

ionic bond: the form of bonding that occurs between two ions when the ions have opposite charges. Sodium cations bond with chloride anions to form common salt (NaCl) when a salty solution is evaporated. Ionic bonds are strong bonds except in the presence of a solvent.

ionise: to break up neutral molecules into oppositely charged ions or to convert atoms into ions by the loss of electrons.

ionisation: a process that creates ions.

irrigation: the application of water to fields to help plants grow during times when natural rainfall is sparse.

isotope: atoms that have the same number of protons in their nucleus, but which have different masses; for example, carbon-12 and carbon-14.

latent heat: the amount of heat that is absorbed or released during the process of changing state between gas, liquid or solid. For example, heat is absorbed when a substance melts and it is released again when the substance solidifies.

latex: (the Latin word for "liquid") a suspension of small polymer particles in water. The rubber that flows from a rubber tree is a natural latex. Some synthetic polymers are made as latexes, allowing polymerisation to take place in water.

lava: the material that flows from a volcano.

limestone: a form of calcium carbonate rock that is often formed of lime mud. Most limestones are light grey and have abundant fossils.

liquid: a form of matter that has a fixed volume but no fixed shape.

lode: a deposit in which a number of veins of a metal found close together.

lustre: the shininess of a substance.

magma: the molten rock that forms a balloon-shaped chamber in the rock below a volcano. It is fed by rock moving upwards from below the crust.

marble: a form of limestone that has been "baked" while deep inside mountains. This has caused the limestone to melt and reform into small interlocking crystals, making marble harder than limestone.

mass: the amount of matter in an object. In everyday use, the word weight is often used to mean mass.

melting point: the temperature at which a substance changes state from a solid to a liquid. It is the same as freezing point.

membrane: a thin flexible sheet. A semipermeable membrane has microscopic holes of a size that will selectively allow some ions and molecules to pass through but hold others back. It thus acts as a kind of sieve.

meniscus: the curved surface of a liquid that forms when it rises in a small bore, or capillary tube. The meniscus is convex (bulges upwards) for mercury and is concave (sags downwards) for water.

metal: a substance with a lustre, the ability to conduct heat and electricity and which is not brittle.

metallic bonding: a kind of bonding in which atoms reside in a "sea" of mobile electrons. This type of bonding allows metals to be good conductors and means that they are not brittle

metamorphic rock: formed either from igneous or sedimentary rocks, by heat and or pressure. Metamorphic rocks form deep inside mountains during periods of mountain building. They result from the remelting of rocks during which process crystals are able to grow. Metamorphic rocks often show signs of banding and partial melting.

micronutrient: an element that the body requires in small amounts. Another term is trace element.

mineral: a solid substance made of just one element or chemical compound. Calcite is a mineral because it consists only of calcium carbonate, halite is a mineral because it contains only sodium chloride, quartz is a mineral because it consists of only silicon dioxide.

mineral acid: an acid that does not contain carbon and that attacks minerals. Hydrochloric, sulphuric and nitric acids are the main mineral acids.

mineral-laden: a solution close to saturation.

mixture: a material that can be separated out into two or more substances using physical means.

molecule: a group of two or more atoms held together by chemical bonds.

monoclinic system: a grouping of crystals that look like double-ended chisel blades.

monomer: a building block of a larger chain molecule ("mono" means one, "mer" means part).

mordant: any chemical that allows dyes to stick to other substances.

native metal: a pure form of a metal, not combined as a compound. Native metal is more common in poorly reactive elements than in those that are very reactive.

neutralisation: the reaction of acids and bases to produce a salt and water. The reaction causes hydrogen from the acid and hydroxide from the base to be changed to water. For

example, hydrochloric acid reacts with sodium hydroxide to form common salt and water. The term is more generally used for any reaction where the pH changes towards 7.0, which is the pH of a neutral solution.

neutron: a particle inside the nucleus of an atom that is neutral and has no charge.

noncombustible: a substance that will not burn.

noble metal: silver, gold, platinum, and mercury. These are the least reactive metals.

nuclear energy: the heat energy produced as part of the changes that take place in the core, or nucleus, of an element's atoms.

nuclear reactions: reactions that occur in the core, or nucleus of an atom.

nutrients: soluble ions that are essential to life.

octane: one of the substances contained in fuel.

ore: a rock containing enough of a useful substance to make mining it worthwhile.

organic acid: an acid containing carbon and hydrogen.

organic substance: a substance that contains carbon.

osmosis: a process where molecules of a liquid solvent move through a membrane (filter) from a region of low concentration to a region of high concentration of solute.

oxidation: a reaction in which the oxidising agent removes electrons. (Note that oxidising agents do not have to contain oxygen.)

oxide: a compound that includes oxygen and one other element.

oxidise: the process of gaining oxygen. This can be part of a controlled chemical reaction, or it can be the result of exposing a substance to the air, where oxidation (a form of corrosion) will occur slowly, perhaps over months or years.

oxidising agent: a substance that removes electrons from another substance (and therefore is itself reduced).

ozone: a form of oxygen whose molecules contain three atoms of oxygen. Ozone is regarded as a beneficial gas when high in the atmosphere because it blocks ultraviolet rays. It is a harmful gas when breathed in, so low level ozone, which is produced as part of city smog, is regarded as a form of pollution. The ozone layer is the uppermost part of the stratosphere.

pan: the name given to a shallow pond of liquid. Pans are mainly used for separating solutions by evaporation.

patina: a surface coating that develops on metals and protects them from further corrosion.

percolate: to move slowly through the pores of a rock.

period: a row in the Periodic Table.

Periodic Table: a chart organising elements by atomic number and chemical properties into groups and periods.

pesticide: any chemical that is designed to control pests (unwanted organisms) that are harmful to plants or animals.

petroleum: a natural mixture of a range of gases, liquids and solids derived from the decomposed remains of plants and animals.

pH: a measure of the hydrogen ion concentration in a liquid. Neutral is pH 7.0; numbers greater than this are alkaline, smaller numbers are acidic.

phosphor: any material that glows when energized by ultraviolet or electron beams such as in fluorescent tubes and cathode ray tubes. Phosphors, such as phosphorus, emit light after the source of excitation is cut off. This is why they glow in the dark. By contrast, fluorescors, such as fluorite, emit light only while they are being excited by ultraviolet light or an electron beam.

photon: a parcel of light energy.

photosynthesis: the process by which plants use the energy of the Sun to make the compounds they need for life. In photosynthesis, six molecules of carbon dioxide from the air combine with six molecules of water, forming one molecule of glucose (sugar) and releasing six molecules of oxygen back into the atmosphere.

pigment: any solid material used to give a liquid a colour.

placer deposit: a kind of ore body made of a sediment that contains fragments of gold ore eroded from a mother lode and transported by rivers and/or ocean currents.

plastic (material): a carbon-based material consisting of long chains (polymers) of simple molecules. The word plastic is commonly restricted to synthetic polymers.

plastic (property): a material is plastic if it can be made to change shape easily. Plastic materials will remain in the new shape. (Compare with elastic, a property where a material goes back to its original shape.)

plating: adding a thin coat of one material to another to make it resistant to corrosion.

playa: a dried-up lake bed that is covered with salt deposits. From the Spanish word for beach.

poison gas: a form of gas that is used intentionally to produce widespread injury and death. (Many gases are poisonous, which is why many chemical reactions are performed in laboratory fume chambers, but they are a byproduct of a reaction and not intended to cause harm.)

polymer: a compound that is made of long chains by combining molecules (called monomers) as repeating units. ("Poly" means many, "mer" means part).

polymerisation: a chemical reaction in which large numbers of similar molecules arrange themselves into large molecules, usually long chains. This process usually happens when there is a suitable catalyst present. For example, ethene reacts to form polythene in the presence of certain catalysts.

porous: a material containing many small holes or cracks. Quite often the pores are connected, and liquids, such as water or oil, can move through them.

precious metal: silver, gold, platinum, iridium, and palladium. Each is prized for its rarity. This category is the equivalent of precious stones, or gemstones, for minerals.

precipitate: tiny solid particles formed as a result of a chemical reaction between two liquids or gases.

preservative: a substance that prevents the natural organic decay processes from occurring. Many substances can be used safely for this purpose, including sulphites and nitrogen gas.

product: a substance produced by a chemical reaction.

protein: molecules that help to build tissue and bone and therefore make new body cells. Proteins contain amino acids.

proton: a positively charged particle in the nucleus of an atom that balances out the charge of the surrounding electrons

pyrite: "mineral of fire". This name comes from the fact that pyrite (iron sulphide) will give off sparks if struck with a stone.

pyrometallurgy: refining a metal from its ore using heat. A blast furnace or smelter is the main equipment used.

radiation: the exchange of energy with the surroundings through the transmission of waves or particles of energy. Radiation is a form of energy transfer that can happen through space; no intervening medium is required (as would be the case for conduction and convection).

radioactive: a material that emits radiation or particles from the nucleus of its atoms.

radioactive decay: a change in a radioactive element due to loss of mass through radiation. For example uranium decays (changes) to lead.

radioisotope: a shortened version of the phrase radioactive isotope.

radiotracer: a radioactive isotope that is added to a stable, nonradioactive material in order to trace how it moves and its concentration.

reaction: the recombination of two substances using parts of each substance to produce new substances.

reactivity: the tendency of a substance to react with other substances. The term is most widely used in comparing the reactivity of metals. Metals are arranged in a reactivity series.

reagent: a starting material for a reaction.

recycling: the reuse of a material to save the time and energy required to extract new material from the Earth and to conserve non-renewable resources.

redox reaction: a reaction that involves reduction and oxidation.

reducing agent: a substance that gives electrons to another substance. Carbon monoxide is a reducing agent when passed over copper oxide, turning it to copper and producing carbon dioxide gas. Similarly, iron oxide is reduced to iron in a blast furnace. Sulphur dioxide is a reducing agent, used for bleaching bread.

reduction: the removal of oxygen from a substance. See also: oxidation.

refining: separating a mixture into the simpler substances of which it is made. In the case of a rock, it means the extraction of the metal that is mixed up in the rock. In the case of oil it means separating out the fractions of which it is made.

refractive index: the property of a transparent material that controls the angle at which total internal reflection will occur. The greater the refractive index, the more reflective the material will be.

resin: natural or synthetic polymers that can be moulded into solid objects or spun into thread.

rust: the corrosion of iron and steel.

saline: a solution in which most of the dissolved matter is sodium chloride (common salt).

salinisation: the concentration of salts, especially sodium chloride, in the upper layers of a soil due to poor methods of irrigation.

salts: compounds, often involving a metal, that are the reaction products of acids and bases. (Note "salt" is also the common word for sodium chloride, common salt or table salt.)

saponification: the term for a reaction between a fat and a base that produces a soap.

saturated: a state where a liquid can hold no more of a substance. If any more of the substance is added, it will not dissolve.

saturated solution: a solution that holds the maximum possible amount of dissolved material. The amount of material in solution varies with the temperature; cold solutions can hold less dissolved solid material than hot solutions. Gases are more soluble in cold liquids than hot liquids.

sediment: material that settles out at the bottom of a liquid when it is still.

semiconductor: a material of intermediate conductivity. Semiconductor devices often use silicon when they are made as part of diodes, transistors or integrated circuits.

semipermeable membrane: a thin (membrane) of material that acts as a fine sieve, allowing small molecules to pass, but holding large molecules back.

silicate: a compound containing silicon and oxygen (known as silica).

sintering: a process that happens at moderately high temperatures in some compounds. Grains begin to fuse together even through they do not melt. The most widespread example of sintering happens during the firing of clays to make ceramics.

slag: a mixture of substances that are waste products of a furnace. Most slags are composed mainly of silicates.

smelting: roasting a substance in order to extract the metal contained in it.

smog: a mixture of smoke and fog. The term is used to describe city fogs in which there is a large proportion of particulate matter (tiny pieces of carbon from exhausts) and also a high concentration of sulphur and nitrogen gases and probably ozone.

soldering: joining together two pieces of metal using solder, an alloy with a low melting point.

solid: a form of matter where a substance has a definite shape.

soluble: a substance that will readily dissolve in a solvent.

solute: the substance that dissolves in a solution (e.g. sodium chloride in salt water).

solution: a mixture of a liquid and at least one other substance (e.g. salt water). Mixtures can be separated out by physical means, for example by evaporation and cooling.

solvent: the main substance in a solution (e.g. water in salt water).

spontaneous combustion: the effect of a very reactive material beginning to oxidise very quickly and bursting into flame.

stable: able to exist without changing into another substance.

stratosphere: the part of the Earth's atmosphere that lies immediately above the region in which clouds form. It occurs between 12 and 50 km above the Earth's surface.

strong acid: an acid that has completely dissociated (ionised) in water. Mineral acids are strong acids.

sublimation: the change of a substance from solid to gas, or vica versa, without going through a liquid phase.

substance: a type of material, including mixtures.

sulphate: a compound that includes sulphur and oxygen, for example, calcium sulphate or gypsum.

sulphide: a sulphur compound that contains no oxygen.

sulphite: a sulphur compound that contains less oxygen than a sulphate.

surface tension: the force that operates on the surface of a liquid, which makes it act as though it were covered with an invisible elastic film.

suspension: tiny particles suspended in a liquid.

synthetic: does not occur naturally, but has to be manufactured.

tarnish: a coating that develops as a result of the reaction between a metal and substances in the air. The most common form of tarnishing is a very thin transparent oxide coating.

thermonuclear reactions: reactions that occur within atoms due to fusion, releasing an immensely concentrated amount of energy.

thermoplastic: a plastic that will soften, can repeatedly be moulded it into shape on heating and will set into the moulded shape as it cools.

thermoset: a plastic that will set into a moulded shape as it cools, but which cannot be made soft by reheating.

titration: a process of dripping one liquid into another in order to find out the amount needed to cause a neutral solution. An indicator is used to signal change.

toxic: poisonous enough to cause death.

translucent: almost transparent.

transmutation: the change of one element into another.

vapour: the gaseous form of a substance that is normally a liquid. For example, water vapour is the gaseous form of liquid water.

vein: a mineral deposit different from, and usually cutting across, the surrounding rocks. Most mineral and metal-bearing veins are deposits filling fractures. The veins were filled by hot, mineral-rich waters rising upwards from liquid volcanic magma. They are important sources of many metals, such as silver and gold, and also minerals such as gemstones. Veins are usually narrow, and were best suited to hand-mining. They are less exploited in the modern machine age.

viscous: slow moving, syrupy. A liquid that has a low viscosity is said to be mobile.

vitreous: glass-like.

volatile: readily forms a gas.

vulcanisation: forming cross-links between polymer chains to increase the strength of the whole polymer. Rubbers are vulcanised using sulphur when making tyres and other strong materials.

weak acid: an acid that has only partly dissociated (ionised) in water. Most organic acids are weak acids.

weather: a term used by Earth scientists and derived from "weathering", meaning to react with water and gases of the environment.

weathering: the slow natural processes that break down rocks and reduce them to small fragments either by mechanical or chemical means.

welding: fusing two pieces of metal together using heat.

X-rays: a form of very short wave radiation.

Index